Texts in Mathematics

Volume 3

Chapters in Mathematics
From π to Pell

Texts in Mathematics Series Editor
Dov Gabbay   dov.gabbay@kcl.ac.uk

Chapters in Mathematics
From π to Pell

Craig Smoryński

ISBN 978-1-84890-053-0

College Publications
Scientific Director: Dov Gabbay
Managing Director: Jane Spurr
Department of Computer Science
King's College London, Strand, London WC2R 2LS, UK

http://www.collegepublications.co.uk

Cover designed by Laraine Welch
Printed by Lightning Source, Milton Keynes, UK

Contents

Dedicatory Poem

Consider, beloved mathematical bibliophage,
The nourishment buried in each successive page.
For the author has taken the wisdom of the ages
And revealed it in various nutritive stages.

—Harry D. Otis

0

Preface

Excuse for this Book

As part of the programme to improve mathematics instruction in the public schools in the city of Chicago, the Chicago Public School System and several of the local universities have joined forces to provide the city's teachers with additional mathematical training. One of the courses offered has been a course in the History of Mathematics for high school teachers. For the Fall of 2008 it was decided that the University of Illinois in Chicago would offer such a course for middle school teachers through the Office of Continuing Education. I was hired to teach it.

The first consideration is a suitable textbook. Most students taking a course in the History of Mathematics in the United States are Mathematics Education majors, preparing to teach mathematics in high schools, and practically all the textbooks in the subject are aimed at them, with a lot of coverage of topics from high school mathematics and the Calculus. These books are hardly suitable for middle school teachers who will have been years away from their high school mathematics and who, for the most part, have had little or no mathematics in college—almost certainly not the standard freshman Calculus course. One text aimed at middle school teachers is

> William P. Berlinghoff and Fernando Q. Gouvêa, *Math through the Ages; A Gentle History for Teachers and Others, Expanded Edition*, Oxton House Publishers and The Mathematical Association of America, 2004.

While I find the title a bit condescending, I suspect middle school teachers would take comfort in it. In any event, the book is less than half the price of most university level history texts and has the organisation I like—a rapid overview followed by a selection of topics to be gone into in greater depth. And it has, in addition to some exercises, a number of projects requiring deeper study. Thus I chose it to be my textbook for the course.

My preference in history is a problem history as in Herbert Meschkowski's volumes *Problemgeschichte der Mathematik*, which, had they been in English, would have been my texts of choice. Berlinghoff and Gouvêa's treatment of topics offers more historical and background material than does Meschkowski, but a lot less mathematical detail. I thus decided to augment their treatment with my own supplementary material on topics I thought would be interesting. The chapters to follow are extended formal write-ups of some of the material I presented to my class. Before discussing these notes, I should add that in teaching this course I was ably assisted by my Teaching Associate Patricia Koch of the Chicago Public School System. The rôle of the Teaching Associate was not clearly defined and she quickly worked out for herself what that rôle was. She gave several presentations in class, discussed possible classroom activities based on material discussed in class, and wrote up and distributed notes of her own.

Contents

As for the topics of the following essays, with the exception of a few brief references to the Calculus, none of them goes beyond high school mathematics. They do go beyond what is currently taught at the middle school, but not beyond what (I think) anyone teaching the more advanced topics of middle school mathematics *should* know. "Should", of course, is a desideratum and not a state of affairs. A vocal minority complained about the algebra, and my colleagues thought the material too advanced. I remain unrepentant in my choice, however. While, for example, the rationale behind Bhāskara's method of solving the Pell equation may be hard to understand, the actual steps in the procedure are doable, not much harder than any number of algorithms we expect (or once expected) middle school students to master without understanding. One can thus skip the explanation of why and work through the how, coming back to the former at a later date should one be interested. In any event, I do not share the utilitarian attitude of those who felt they should only learn what they would be covering in their own teaching. That is no way to develop a perspective on mathematics or to prepare for the inevitable changes in curriculum that will come.

The opening chapter on approximating π mixes history, classroom activities, and some instructions on using the calculator as well as a first example of the limitations of the unprogrammed calculator. I did not follow up on discussing what could be done in the classroom as Ms. Koch was much better at that than I was. The second chapter is the exception that proved this rule. The point was not to prove Napoleon's Theorem so much as to emphasise that many famous people have connexions with mathematics, a fact that could be exploited in the classroom to relate mathematics to other subjects being studied.

Chapter 3 is presented as a follow-up to Chapter 2 in that it derives the results used in the proof of Napoleon's Theorem. But its real purpose is to present something from Euclid, which one simply must do in a course on the History of Mathematics. That Euclid and Ptolemy provide the raw material used in the proof in Chapter 2 determined which selections from Greek authors to include.

Chapters 1 through 3 all concern geometry, whence I have grouped them together as Part I on geometry. Part II dealing with computation is undoubtedly more massive than it should be and some explanation is required. It had never been my intention to go so deeply into any topics as I did in this Part. What happened was that while giving a brief outline of the history of mathematics in India, I mentioned, as Berlinghoff and Gouvêa report, that Brahmagupta solved the equation

$$X^2 - 92Y^2 = 1,$$

and stated that anyone who could solve this equation in under a year was a mathematician. The class asked how he did it and I, as an enthusiast as regards the Pell equation, needed no further encouragement. The write-up, of course, goes beyond what I could present to my students, but I did distribute the full chapter to them. Even the lightest skimming of the pages shows that the Pell equation is a matter of some depth.

While I was lecturing on the Pell equation, one student asked about the Euclidean Algorithm, which became my next topic. After this I thought to return to the Pell equation using the Euclidean Algorithm as a simpler, if less historical approach. Calculating the solutions by hand using this algorithm is a sure way to commit errors, both because of the increasing sizes of the numbers involved and all the minus signs. I very quickly wrote the programs of Chapter 9 solving the Pell equation on the computer but never presented them in class. I had in mind to move the program over to the calculator or, at least, to LOGO—which move requires dealing exactly with integers. In the meantime, however, I could give an example or two calculated by hand. The simplest of these concerns the golden ratio, any discussion of which necessarily involves the Fibonacci numbers.

The Fibonacci numbers form, I think, a good choice of topic for a course on the History of Mathematics for middle school teachers. They are historical and they can be fun. They can also be used to introduce some other topics of possible interest—induction, recursion, recreational mathematics, matrices, etc. Chapter 6 is an imperfect representation of what was covered in class. After verifying that induction is a topic for high school teachers and not for middle school teachers, I skipped that material, and the discussion of matrices was a later addition to my notes, suggested to me by the class's interest in taking a course on calculators to be offered in the spring. I cannot remember whether I distributed this latter material or not. I may have decided, after writing it up, that it was a bit too novel to be of any use to them. Additionally, in class I discussed the Towers of Hanoi puzzle in connexion with the mention of Édouard Lucas. It had always been my intention to give a lecture on recreational mathematics, but the semester ended before I got round to doing so. I might also add that Ms. Koch rounded out the discussion with classroom activities involving Fibonacci numbers and some words on the various appearances of the numbers in nature.

The straight historical discussions of the course are not represented here except for a bit of Chapter 7, which expands slightly on what I had to say about the 19th century. The most accessible 19th century mathematics is the work on the foundations of mathematics and the discovery at the end of the century of the paradoxes of set theory. While the foundational work of the 19th century itself is beyond the level of preparation of the average middle school teacher, the problems it addresses are readily exemplified. Thus I presented such paradoxes of the infinite as infinite decimal expansions (e.g., $1 = .999...$) and the Grandi series. The detailed discussion of Euler, the Fibonacci numbers, and partial fraction expansions was, of course, too algebraically intense for in class presentation and was only added later for the amusement of the more advanced students.

The initial parts of Chapter 8 on finding square roots is the last of Part II that I actually covered in class and was the last set of notes I distributed. My main concern was the algorithm that used to be taught in elementary school for finding square roots. While the history of algorithms for finding square roots is interesting in itself, I thought it might be of added interest as a whisper of a hint of the existence of a history of mathematics education. Also, it does supply one with an example where one actually works in a base other than 10. I don't know that anyone took the bait, and I might have scared them off when, at the last minute I decided to depart from my notes in class and find $\sqrt{2}$ instead of $\sqrt{189574}$. Needless to say, there is a slight complication in the case chosen that I had forgotten about and I got hopelessly confused. Not being one to think on my feet in front of an audience, I relegated the corrections to the notes. This was one of the reasons why I decided to program the calculator to perform the algorithm for me. The program was, however, appallingly slow, as reported here in Chapter 8, and I never got round to presenting it in class or even distributing it to my students.

Before saying anything about Chapter 9, I should pause to say a few words about the calculator and the computer. I have two calculators in my collection that were standard issue at two of the universities I've taught at. Both are manufactured by Texas Instruments and it appears most colleges and universities in the United States have chosen *TI* calculators as standard, the predominant choice being some variant of the *TI-83* (the *TI-83 Plus* or *TI-84*) for pre-Calculus and the *TI-89* for Calculus and beyond. I have the *TI-83 Plus* and the *TI-85*, which latter is now obsolete but was formerly a solid choice for the advanced courses. Because of the widespread use of the *TI-83* family, I have seen fit to base my exposition on the *TI-83 Plus* whenever I was compelled to use a calculator. The programming language for this calculator, popularly called TI-BASIC, is a variant of BASIC and the programs should be readable enough that one using a different calculator should have no difficulty in adapting them to his calculator.

I have also used the computer on a number of occasions. Insofar as LOGO is a very simple, very powerful language originally designed for elementary school students, I have included a number of programs in that language. It turns out that it is not much taught any more—I know not why—but I have kept the programs anyway because they are simple and should be easy enough to read even by one with no prior familiarity with the language.

In Chapter 9, however, I use SCHEME. The move to the computer from the calculator was dictated by the slowness of the square root program of Chapter 8 on the calculator, and the fact that programming on the computer allows one to create files, even formatted tables. I started working in LOGO on the *iMac*, but not all compatibility issues between software written for OS X and the new Intel chip seem to have been worked out, and the LOGO editor in the implementation I was using was acting strangely. Thus I switched over to SCHEME, which had the additional advantage of dealing exactly with integers. I did not have to program the necessary added precision via lists as I did on the *TI-83 Plus* or would have had to have done in LOGO. Writing programs in SCHEME is not quite as easy as in LOGO, but the programs, once written, are just as easy to read by anyone familiar with another language and, should he prefer BASIC or PASCAL or C augmented by numerous plusses, he should have no difficulty in translating them. My preference for SCHEME was based on i) its affinity to LOGO, and ii) the fact that it can be downloaded for free.

As to the contents of Chapter 9, they are these: First there is a version of the square root program of Chapter 8 written in SCHEME, and a table of square roots of the integers from 0 to 100 to 50 decimals. Then there is the program, again written in SCHEME, to solve the Pell equation by application of the Euclidean Algorithm, together with several computer generated and formatted tables related thereto. This chapter is not really for the middle school teachers, and is not very historical, but is, I think, moderately interesting—and it was definitely fun doing the programming.

Chapter 10, written after the course was finished, concerns the history of tables, a subject that seemed an appropriate close to Part II after all the tables of Chapter 9.

One of my students showed me his middle school mathematics textbook and I noticed it contained material on probability. This was a surprise to me. It was not a subject for middle school when I was of middle school age. There was a chapter on the topic in our Intermediate Algebra book when I was in high school, and I read it with interest, but the material was not covered in class—and, academically, my high school was perhaps a bit better than average. Thus, I would not have thought of discussing probability and some elements of its history with middle school teachers had I not seen it in my student's textbook. I devoted the last week of the course to probability. The material is not presented here. I only started writing up my presentation after the course was finished, and without the constraints of the course, the material kept expanding. After 4 chapters, with more in sight, I realised that in order to keep the book down to manageable size, I would have to reserve these chapters for presentation elsewhere.

Parting Comment

I hesitate to advertise the present book as a supplement for a history course aimed at middle school teachers, although that was the point of departure in my setting out to write it. It has evolved away from the narrow confines of my course. I would nonetheless recommend the book to the prospective teacher of the more advanced topics of middle school mathematics or of high school mathematics, or, indeed the high school graduate who has a liking for mathematics and might like some more, presented a bit differently. It is not light reading, the selection of topics is diverse and not directly aimed at one's upcoming college courses, but I think the reader will find a careful study of the book rewarding and perhaps even fun.

Acknowledgments

Three individuals must be thanked: John Baldwin, who got me involved in teaching the course in the first place; Patricia Koch, whose classroom contributions helped keep the course from deviating off-course the way this book did; and Eckart Menzler-Trott whose comments and advice, especially in the material that has now migrated over to my forthcoming volume on probability, have been most valuable.

Part I

Geometry

1

The Value of π

Mathematicians like to start the history of mathematics with the Greeks and the origin of proofs. But there were some major discoveries made in the pre-Greek empirical period that merit consideration. There were of course the basic problems of numeration—representing numbers—and computation with the numerals. These were solved in various ways in different cultures. *How* is, of course, unknown—and not very interesting because the question does not seem amenable to mathematical analysis. Two questions that are of interest that can profitably be explored, though never answered definitively, are the origins of the Pythagorean Theorem and π.

The Pythagorean Theorem would not seem to be an empirical discovery. I mean, what would possess one to find the areas of the squares on the sides of a right triangle, compare them, and notice a relationship? It is the sort of theorem that would be suggested by a proof, perhaps one of those cited in Berlinghoff and Gouvêa. (On the other hand, one could discover Pythagorean number triples like 3, 4, 5 and at some point observe that the 3-4-5 triangle is a right triangle...)

Far more interesting is π and the question related to its discovery. How did one first realise that

i. the ratio of the circumference (C) of a circle to its diameter (d) or radius (r) is constant,

$$\frac{C}{d} = \frac{C}{2r} = \text{constant};$$

ii. the ratio of the area (A) of a circle to the square of its diameter is constant,

$$\frac{A}{d^2} = \frac{A}{(2r)^2} = \text{constant; and}$$

iii. these constants are related:

$$\frac{C}{2r} = 4 \cdot \frac{A}{(2r)^2} = \frac{A}{r^2}?$$

The history textbooks report on various values of π used in different cultures. Do they say whether the given constant is the constant of i, or ii, or a common value iii for both? For example, they report that the ancient Jews used the value 3 for π by referring to a Biblical passage (1 *Kings* 7:23) in which a circular pool measures 10 cubits across and 30 cubits around. Thus 3 is clearly the value of π associated with the circumference of the circle. No mention is made of the area of the circle. Is it reported anywhere that the ancient Jews knew fact ii that the area of the circle is proportional to the square of its radius, or that this ratio is the same π they used 3 as an approximation of? I suppose I should also ask how they arrived at the number 3 as an approximation of the ratio of the circumference to the diameter, but the solution with a tape measure is so obvious that any other explanation would appear far fetched. Nonetheless, I note that if one approximates the circumference of a circle by the perimeter of its inscribed hexagon one gets the value

$$\frac{C}{2r} = \frac{6}{2 \cdot 1} = 3,$$

as one sees by a glance at FIGURE 1.

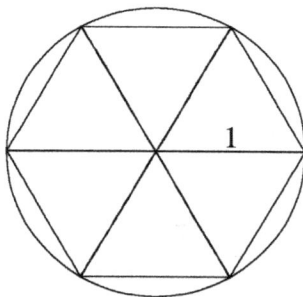

FIGURE 1.

More intriguing is the question of the Egyptian value of π:

$$\pi \approx 4 \cdot \left(\frac{8}{9}\right)^2 \approx 3.1604930827,$$

or, more accurately rounded down, 3.16. Geometrically this is an easy number to arrive at. In *The Crest of the Peacock*, a nice popular book on mathematics in non-Western cultures, George Gheverghese Joseph offers a couple of explanations he found in the literature that are simple enough that elementary school children could grasp them.

  For the first of these, one notices that an octagon is a good approximation to the circle. So take a 9-by-9 square and inscribe an octagon as in FIGURE 2 and determine the area of the octagon. It is just the area of the square, minus the areas of the 4 triangles, i.e.,

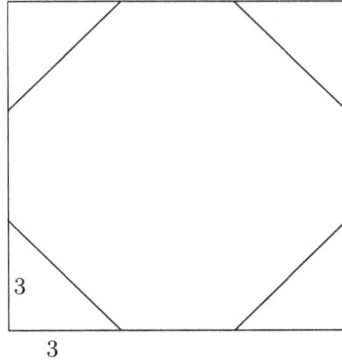

FIGURE 2.

$$9 \cdot 9 - 4 \cdot \frac{1}{2} \cdot 3 \cdot 3 = 81 - 18 = 63 \approx 64 = 8^2.$$

If we take this to approximate the area of the circle of diameter 9 inscribed in the octagon, then the value of π is approximately

$$\frac{8^2}{(9/2)^2} = 2^2 \cdot \frac{8^2}{9^2} = 4 \cdot \left(\frac{8}{9}\right)^2.$$

This explanation is plausible in that the figure drawn appears, with the hieratic numeral 9, in the Rhind Mathematical Papyrus[1] in problem 48, near a couple of problems dealing with the area of the circle. Joseph says, however,

> There is something rather contrived and unconvincing about this explanation, for it assumes an algebraic mode of reasoning which is not immediately apparent in Egyptian mathematics.[2]

Instead Joseph refers to the popularity of geometric designs in ancient Egypt. He cites

---

[1] Named after Alexander Henry Rhind (1833 - 1863), who purchased it in Egypt and willed it to the British Museum. It is the most important single document on mathematics from ancient Egypt still extant. Dating to 1650 B.C., it is a copy made by a scribe Ahmose (Ahmes in older transcriptions) of an earlier work and is largely a collection of worked problems for the training of scribes.

[2] George Gheverghese Joseph, *The Crest of the Peacock: Non-European Roots of Mathematics*, Penguin Books, London, 1991, p. 83.

P. Gerdes, "Three alternative methods of obtaining the ancient Egyptian formula for the area of a circle", *Historia Mathematica*[3] 12 (1985), pp. 261 - 268.

Joseph presents a diagram based on Gerdes which I reproduce in an especially artistic manner as FIGURE 3 which shows a circle of diameter 9 being closely

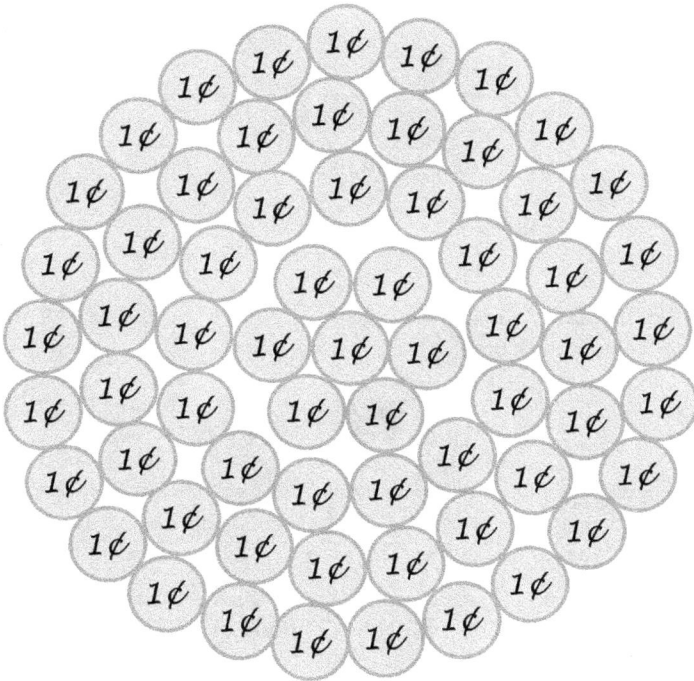

FIGURE 3.

covered by 64 discs of diameter 1. This could be demonstrated to elementary school students with 64 pennies (or buttons) on an overhead projector. Indeed, one could start with an 8-by-8 arrangement of the pennies and then slide them into place to show that the areas of a square of side 8 and a circle of diameter 9 are very close.

---

[3] *Historia Mathematica* is the main journal devoted to the history of mathematics. For more general history of science there are also *Archive for History of Exact Sciences*, *Isis*, and *Osiris*. There are also interesting articles of an historical nature in *The American Mathematical Monthly*. Next time you visit the university library you should take a look at these journals and see if any of their articles piques your interest. I should think the articles in the *Monthly*, in having more mathematics, would be particularly interesting.

My favoured approach, which could be presented as a worksheet activity, is to use a circle of diameter 9 inscribed in a 9-by-9 square, with grid showing, as in FIGURE 4. The students could then approximate the area of the circle by

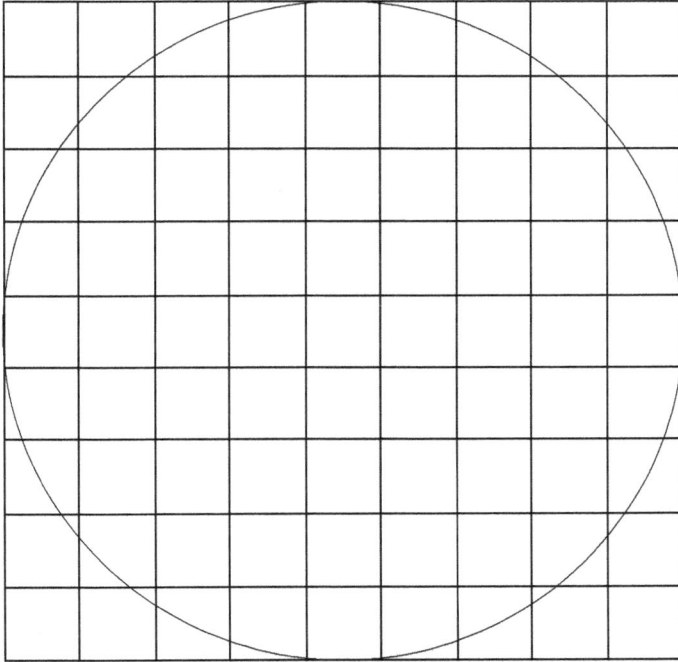

FIGURE 4.

answering a series of simple questions:

1. How many of these squares are nearly completely covered by the circle?

2. How many squares are approximately half covered by the circle?

3. Multiply this number by 1/2 and add it to answer 1.

4. How many squares are approximately 3/4 covered by the circle?

5. Multiply this number by 3/4 and add it to answer 3.

6. What approximation to π does all this yield?

The Chinese used several values for π. In fact, in the *Shùshū jiǔzhāng*, or *Mathematical Treatise in Nine Sections*, of 1247 A.D., Qín Jiǔsháo (*c.* 1202 - *c.*1261) used three values of π:

the old value   3

Zhāng Héng's value $\sqrt{10}$

the accurate value $\frac{22}{7}$

We have already explained how one might obtain the old value. Zhāng's value
is quickly derived from the Egyptian value by squaring it and noting that it is
slightly smaller than 10, whence $\sqrt{10}$ is another approximation:

$$\left(4 \cdot \left(\frac{8}{9}\right)^2\right)^2 = \frac{16 \cdot 8^4}{9^4} = \frac{65536}{6561} \approx 9.9887\ldots \approx 10.$$

This is unlikely to be the derivation of the value as $\sqrt{10}$ is no handier to
work with than the more accurate Egyptian value. I tend to favour using a
tape measure: A circle 1 foot in diameter will be just under 3 feet 2 inches in
circumference, so

$$3 + \frac{2}{12} = 3 + \frac{1}{6} = \frac{19}{6}$$

makes a good approximation. Squaring this gives

$$\left(\frac{19}{6}\right)^2 = \frac{361}{36} \approx \frac{360}{36} = 10$$

and again $\sqrt{10}$ is an approximation to π. It is, in fact, more accurate than $\frac{19}{6}$.

$\sqrt{10}$ also served as an approximation to π among Indian mathematicians.
Hermann Hankel (1839 - 1873) suggested that the early Indian mathematicians
may have calculated that the perimeters of the regular 12-, 24-, 48- and 96-gons
of diameter 10 were $\sqrt{965}$, $\sqrt{981}$, $\sqrt{986}$, $\sqrt{987}$, respectively, and suspected that
they approached $\sqrt{1000}$ in the limit, leading to a value

$$\pi = \frac{\sqrt{1000}}{10} = \sqrt{10}.\text{[4]}$$

The "accurate value" $\frac{22}{7}$ was first obtained by Archimedes (c. 287 B.C. -
212 B.C.) by just such an appeal to the 96-gon and thus some credence is lent
to Hankel's conjecture. It is not too hard to determine the recurrence for the
perimeter of a regular $k \cdot 2^n$-gon for fixed k and fixed diameter. Starting with
the hexagon ($k = 6$), it requires only 4 doublings to obtain the result for the
96-gon.

Mention of Archimedes brings us out of the empirical period and deep into
the Greek. It is not clear who knew what when among the Greeks. Aristotle
(384 B.C. - 322 B.C.) refers to Antiphon (*fl.* 2nd half of the 5th century B.C.)
who is said to have started with an equilateral triangle inscribed in a circle
and successively doubled the number of sides eventually exhausting the circle.

---

[4] Hermann Hankel, *Zur Geschichte der Mathematik in Alterthum und Mittelalter,*
Verlag von B.G. Teubner, Leipzig, 1874, pp. 216 - 217.

Whether this meant he thought he could obtain the circle at some finite stage, that he could do the doubling infinitely many times to represent the circle as a polygon with infinitely many sides, or that he was merely getting a sequence of successively closer approximations to the circle is a matter of debate. By the time of Euclid, *c.* 300 B.C., one had a good enough handle on approximations that one could prove theorems. In *The Elements* Euclid showed the ratio of the areas of two circles to equal the ratios of the squares on their diameters:

$$\frac{A_1}{A_2} = \frac{(d_1)^2}{(d_2)^2}.$$

Although he also proved laws of proportions like

$$\frac{a}{b} = \frac{c}{d} \Rightarrow \frac{a}{c} = \frac{b}{d},$$

he did not explicitly draw the conclusion ii that π exists:

$$\frac{A_1}{(d_1)^2} = \frac{A_2}{(d_2)^2}.$$

Presumably this is because of the limited Greek notion of number. Today, however, we have no qualms about drawing the conclusion.

Similarly, accepting that the circumference of a circle is greater than the perimeter of any inscribed polygon and less than that of a circumscribed polygon, the same method of proof establishes that the ratio of the circumference of two circles is the same as that of their diameters:

$$\frac{C_1}{C_2} = \frac{d_1}{d_2},$$

and thus the existence of π in the sense i:

$$\frac{C_1}{d_1} = \frac{C_2}{d_2}.$$

In "Measurement of the Circle" (*c.* 250 B.C.), Archimedes establishes as his first proposition that the area of a circle equals the area of a right triangle with sides equal respectively to the radius and circumference of the circle. From this it follows immediately that the two definitions of π are equivalent, i.e., iii holds. The rest of "Measurement of the Circle" is devoted to approximating π. He shows, by comparing the circumference of the circle with the perimeters of inscribed and circumscribed regular 96-gons, that

$$3\frac{10}{71} < \pi < 3\frac{1}{7} = \frac{22}{7}.$$

Actually reading Archimedes on this point is not very rewarding. The ideas are obscured by the manner of expressing them. It is really quite a simple matter. A couple of applications of the Pythagorean Theorem allow one to express the the side of (say) the inscribed regular $2m$-gon in terms of the side of the inscribed regular m-gon. From there the perimeter as an approximation to the circumference of the circle and an approximation of π are readily calculated.

Suppose we have a circle of radius r centered at O, and a chord AB of length x representing a side of an inscribed regular m-gon. To determine the length z of the side of an inscribed regular $2m$-gon, we draw the bisector of AB as in FIGURE 5. Now OBD is a right triangle, whence we can use the Pythagorean

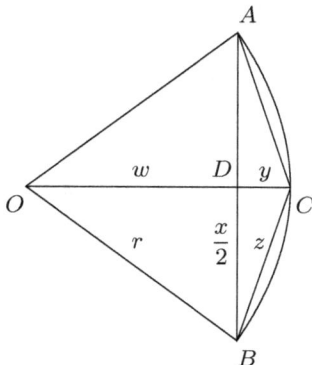

FIGURE 5.

Theorem to conclude

$$w^2 = r^2 - \left(\frac{x}{2}\right)^2 = r^2 - \frac{x^2}{4} = \frac{4r^2 - x^2}{4}.$$

BDC is a right triangle and the Pythagorean Theorem again applies:

$$z^2 = y^2 + \left(\frac{x}{2}\right)^2 = (r - w)^2 + \left(\frac{x}{2}\right)^2$$

$$= r^2 - 2rw + w^2 + \frac{x^2}{4}$$

$$= r^2 - 2r\frac{\sqrt{4r^2 - x^2}}{2} + \frac{4r^2 - x^2}{4} + \frac{x^2}{4}$$

$$= 2r^2 - r\sqrt{4r^2 - x^2},$$

and we have

$$z = \sqrt{2r^2 - r\sqrt{4r^2 - x^2}}.$$

What is left now is just a matter of numerical work, which today we would do by calculator. For me that means my trusty *TI-83 Plus*. The sequential mode of this calculator is designed to handle such recurrences as that iterating the passage from x to z. It does have its quirks however.

Preparatory to dragging out the calculator, let us agree to start with the regular hexagon and double the sides from there *á la* Archimedes and Hankel's Indians. As for the choice of radius, the most convenient one is $r = 1$, but in discussing Hankel's explanation of the use of $\sqrt{10}$ as an estimate for π, the value $r = 5$ is more natural. Thus, we should think of working with a stored value of R for the radius in our calculator. The recursion for the length $s(n)$ of a side of the regular $6 \cdot 2^n$-gon obtained by doubling the number of sides n times is

$$s(0) = r$$

$$s(n+1) = \sqrt{2r^2 - r\sqrt{4r^2 - s(n)^2}}.$$

Now store your value of r in the variable R on the calculator, then press the MODE button on the calculator, scroll down to the 4th row, select Seq and press ENTER. Then press the Y= button to enter the equation editor and enter the following:

0	for	nMin=
$\sqrt{(2 * R^2 - R * \sqrt{(4 * R^2 - u(n-1)^2)})}$	for	$u(n) =$
R	for	$u(n$Min$)=$

If you now hit ENTER or exit the screen and return, you will find an annoying quirk of the *TI-83 Plus*: The value you stored for R and not the variable occurs as the value of $u(n$Min$)$. Thus, as you switch back and forth between $r = 1$ and $r = 5$, you must not only store the new value in R, but also reset the value of $u(n$Min$)$.

You might as well also enter $6 * 2^n$ for $v(n)$ to calculate the number of sides of the polygon after n doublings. You can now do your explorations. First, create a list by entering the LIST menu and choosing seq:

$$\mathsf{seq}(n, n, 0, 12) \to \mathsf{L_1}.$$

This just lists $0, 1, 2, \ldots, 12$, the total number of doublings of sides considered. Then enter

$$u(0, 12) * v(0, 12) \to \mathsf{L_2}.$$

This will give us a list of the perimeters of the various inscribed polygons. [$u(n)$ gives the n-value of the sequence u. $u(n, m)$ will produce a list of values $u(k)$ for $k = n, n+1, \ldots, m$.] Entering

$$\mathsf{L_2}/(2 * \mathsf{R}) \to \mathsf{L_3}$$

will give a list of the approximations to π obtained by the iterated doublings of the number of sides. You might want to consider the 5th entry (4 doublings, i.e., 96 sides) and verify that it lies between $3\frac{10}{71}$ and $3\frac{1}{7}$.[5] It is correct to 3 decimals. The 9th entry (8 doublings, i.e., 384 sides) is Āryabhata's estimate (*c.* 500 A.D.) and is correct to 5 decimal places. The 13th entry (12 doublings, i.e., 24576 sides) is Zŭ Chōngzhī's (*c.* 429 - *c.* 500) estimate (*c.* 480 A.D.) and

---

[5] This, of course, is not the same as showing π to lie between the two values, but it is comforting nonetheless.

is correct to 7 decimals. In 1429, the Persian al-Kāshī (d. 1429) used a polygon with $6*2^{27}$ sides to obtain a value of π correct to 16 decimals, and a century and a half later in 1579 François Viète (1540 - 1603) used a polygon with 393216 sides (16 doublings) to obtain 9 decimals.

In theory what I have just said is true. In practice, however, we start running into the limitations of the calculator. Obviously, al-Kashi's 16 decimals are too many for the calculator to display. One might hope, considering that the display of most of the approximations to π is given to 9 decimals, that one might get all but maybe the last couple of digits of Viète's value. One doesn't: the value given by calculating u(16)*v(16)/2 is correct only to 3 places. (And if one tries 17 doublings the result is correct only to two decimals!) Even the value obtained for Zǔ's estimate is correct only to 6 places. The accumulation of rounding errors is devastating.[6] Proper exploration of the method clearly requires one to program some extra precision into one's calculations or to transfer the problem to a dedicated mathematics program such as MAPLE or MATHEMATICA.

For a small number of doublings, however, one should be safe enough and I suggest the exercise of carrying out the above computations for R = 5, and using one more list:
$$\mathsf{L_2}^2 \to \mathsf{L_4}$$
and comparing these values for the squares of the perimeters of the polygons with Hankel's 965, 981, 986, 987. What do you notice for 5, 6, 7, ... doublings? How plausible is Hankel's explanation of the use of $\sqrt{10}$ as a value of π?

That Euclid, Zǔ, al-Kashi, and Viète used the method of inscribing polygons is a "fact" I gleaned from *The Crest of the Peacock*. It is certainly the conceptually simplest method and, in Zǔ's case, we know Zǔ would have been familiar with Liú Huī's ($fl.\ c.$ 250 A.D.) appeal to such constructions to obtain various estimates of π. However, according to Lǐ and Dù, there is not enough material in Zǔ's surviving works to allow one to state definitely what method he used. I leave it to the reader to check the literature for the methods used by the others.

Incidentally, another value Zǔ gave, correct to 6 decimals, is $\frac{355}{113}$, an easily memorised value not discovered in Europe until near the end of the 16th century. This value is the upper bound analogous to Archimedes's $\frac{22}{7}$ for one more doubling to 192 of the number of sides of the inscribed regular polygon:

$$\frac{3549}{1130} < \mathsf{L_3}(6) \approx 3.14145\ldots < \pi < 3.1415292\ldots \approx \frac{3550}{1130} = \frac{355}{113}.$$

It is also the next convergent after $\frac{22}{7}$ of the continued fraction expansion of π and nowadays is obtained via a lot of theory and not much calculation.

---

[6] Lǐ Yǎn and Dù Shíràn, (John N. Crossley and Anthony W.-C. Lun, translators), *Chinese Mathematics; A Concise History*, Oxford University Press, Oxford, 1987, p. 84, note that, if Zǔ indeed used this method, "then Zǔ Chōngzhī had to perform very complicated operations on nine-digit numbers more than one hundred times (including taking square roots)". It is thus easy to imagine the accumulation and growth of rounding errors.

2

Emperor of Mathematics

Many people whose names are more commonly associated with non-mathematical pursuits yet have their place in mathematics. The famous French circumnavigator Louis Antoine de Bougainville wrote a Calculus textbook. The first president of Ireland, Eamon de Valera, had been a mathematics teacher, as was at one time the actor Sam Jaffe. The actress Danica McKellar co-authored a paper on applied mathematics and additionally wrote a couple of popular accounts of mathematics aimed at attracting more girls to the subject. The American general and president Ulysses S. Grant did not become a mathematics teacher as he had planned, having been sidetracked into a more traditional military career. President Garfield came up with his own proof of the Pythagorean Theorem. The list of examples could be multiplied endlessly. The member of the list I wish to discuss here is Napoleon Bonaparte (1769 - 1821).

That Napoleon was a patron of science is well-known. Any account of Egyptology will report that his expeditions to Egypt were as much scientific as military. His background and interest in mathematics and the attribution to him of a theorem of geometry are less reported. Indeed, a quick check of the literature will find contrary opinions on Napoleon's mathematical ability and the possible authorship of the result. While it is true that there is no direct evidence linking Napoleon to the theorem named after him, he certainly had the background to enable him both to discover and prove the result. His mathematical abilities were attested to by his teachers who felt them strong enough to recommend he join the artillery, where such skills were needed. And it is reported that while in Italy he studied Lorenzo Mascheroni's (1750 - 1800) *Geometria del Compasso* (1797) with the author and later, on returning to Paris, reported on it to the French mathematicians Joseph Louis Lagrange (1736 - 1813) and Pierre Simon de Laplace (1749 - 1827).

Napoleon's Theorem is not a cornerstone of mathematics. It does not appear extremely useful and might best be assigned to what used to be called recreational mathematics. It certainly does not appear in every geometry textbook, something I can testify to from having spent a fruitless afternoon a few years ago searching for it in a decent university library. Nowadays, with the

Internet, it is much easier to find information on such things. A quick search yields pages on the Web where the result is carefully stated and several proofs are presented. Before we discuss any proofs here, we must, of course, have a statement of the result:

1 Theorem (Napoleon's Theorem). *Let a triangle ABC be given and erect on the exterior of each side an equilateral triangle with the side as a base. The triangle obtained by connecting the centres of the equilateral triangles is itself equilateral.*

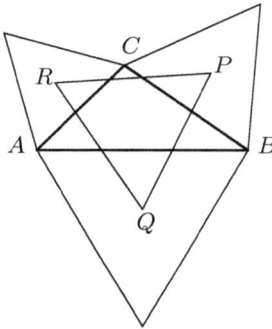

FIGURE 1.

One might ask first how on earth one would discover such a thing. Well, Napoleon was Italian, born on the island of Corsica, went on to glory in France, and had an interest in mathematics. It would be most natural for him to have taken a special interest in the works of French and Italian mathematicians, and there is a closely related problem that had been studied by French and Italian mathematicians a couple of centuries earlier.

In the early days of the Calculus, before Newton (1642 - 1727) and Leibniz (1646 - 1716) came up with the bag of algorithms that gave the subject its name, mathematicians devised a number of techniques to find areas and tangents, and solve related problems. One of these was Pierre de Fermat (*c.* 1601 - 1665), a lawyer by trade who is remembered today as one of the great mathematicians. A contemporary of Rene Descartes (1596 - 1650) and Blaise Pascal (1623 - 1662), he and Descartes independently invented analytic geometry, while, in their correspondence, Fermat and Pascal pioneered the newly developing field of probability. Fermat also devised a technique for solving maximum-minimum problems. To demonstrate the power of his method, he threw out the following challenge to his contemporaries:

2 Problem. Given a triangle ABC, all the angles of which are less than $120°$, find the point F in the interior for which the total distance $AF + BF + CF$ is minimised.

It turns out that if one of the angles is 120° or greater, the minimisation problem is solved by the vertex of the large angle, i.e,. the vertex connecting the two shortest sides, and is not interior to the triangle, hence the restriction.

Fermat's problem was solved by Galileo's (1564 - 1642) student Evangelista Torricelli (1608 - 1647), who found F by erecting the equilateral triangles on the sides of the given triangle ABC. He showed the point F to be the point of intersection of the three circles circumscribing the equilateral triangles. The picture one draws is very similar to FIGURE 1 and, if one draws the triangle ABC and its attached equilateral triangles and adds the points at the centres of these latter triangles before drawing the circumscribing circles centred on them, one needs only connect the dots to arrive at FIGURE 1. It shouldn't take too long staring at such a figure to form the conjecture that the distances between these centres are equal.

A little more about the Fermat point F before discussing Napoleon's Theorem. Another contemporary of Fermat's was Bonaventura Cavalieri (1598(?) - 1647), who also looked into the problem and proved that F was the unique point in the interior for which the angles $\angle AFB, \angle AFC, \angle BFC$ are all 120°. And a century later Thomas Simpson (1710 - 1761) characterised F as the point of intersection of the three lines connecting the vertices of the triangle ABC with the far vertices of the equilateral triangles opposite them; moreover, *Simpson's lines* are all of the same length and the 6 angles formed by these lines at F are each 60°.

Now, Simpson's result holds even when the restriction that all angles in ABC be less than 120° is dropped. If an angle is greater than 120°, however, the point F is exterior to the triangle. Nonetheless, F is still the point of intersection of the three circumscribing circles.

I mention all of this because there is a close relation between Napoleon's Theorem and the more purely geometric characterisations of F (i.e., those other than that it solves the minimisation problem): Napoleon's Theorem can be derived as a corollary from the results on the Fermat point. And I would expect all of this to be intelligible to a middle school student. I base this on the knowledge that the results of Cavalieri and Simpson were rediscovered and proven by a 13-year old. He was, of course, not just any 13-year old, but a 13-year old genius. I would not expect the average 13-year old to be able to discover the result, much less come up with his own proof.

The 13-year old in question was Gerhard Gentzen (1909 - 1945), who grew up to be one of the 20th century mathematicians responsible for transforming mathematical logic from an application into a tool. Gentzen had been studying geometry in school when he turned his attention to the diagram underlying FIGURE 1 in which P, Q, R have not yet been added. To this he added Simpson's lines, noticed they intersected in a point, and continued from there. Now the question of what motivated young Gentzen to consider such a configuration in the first place might arise, and there is a natural enough answer: Gentzen had been studying geometry in school and presumably had been taught Euclid's proof of the Pythagorean Theorem with the famous "stiltwalking" diagram of FIGURE 2 in which Euclid drew a perpendicular line (CJ) from the right angle

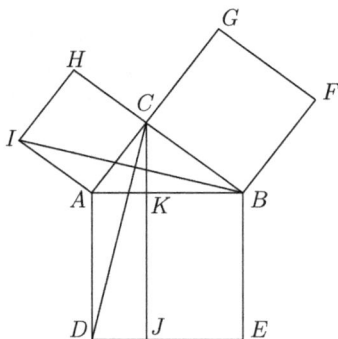

FIGURE 2.

vertex of the right triangle to the far side of the square opposite, as well as lines (CD and BI) from the vertices to the far vertices of their opposing squares. Gentzen may have decided to see what happened if he replaced the squares by equilateral triangles—the easiest similar polygons to affix to the triangle ABC using ruler and compass. Now there are no far sides of the equilateral triangles to connect the vertices of ABC to, but there are the vertices. Thus, he would have discovered Simpson's lines and the slightest care in drawing the figure would have revealed their common point of intersection, and a conjecture for him to prove.

I don't propose to prove the results on Simpson's lines and the Fermat point here. I refer the reader to Paul J. Nahin's book *When Least is Best* (Princeton University Press, Princeton, 2004, pp. 279 - 286) for a nice exposition on the subject and to my own essay, "Gentzen and geometry", which I included as an appendix to the English translation of Eckart Menzler-Trott's biography of Gerhard Gentzen, *Logic's Lost Genius; The Life of Gerhard Gentzen* (American Mathematical Society, Providence (Rhode Island), 2007). Therein I recount Gentzen's proofs of various results about the Fermat point and Simpson's lines and show additionally how the methods he used yield Napoleon's Theorem as a corollary.

As mentioned earlier, a statement of Napoleon's Theorem is usually missing from geometry textbooks. It follows that the proof will also be missing. However, in recent years, such things have been popping up on the Internet and it is easy to find sites featuring a number of proofs, both synthetic and analytic. What one means here by *synthetic* is a proof, á la Euclid, using strictly geometric methods; an *analytic* proof is a computational one. One could choose a coordinate system and calculate the coordinates, as in analytic geometry, or calculate via some other means such as trigonometric ones. Indeed, since we are dealing with triangles, appeal to trigonometry and some basic trigonometrical laws would seem to be the obvious and appropriate approach to the problem. For students who know trigonometry, it offers the most reproducible proof: one sets about calculating the lengths of the sides PQ, QR, PR of the new triangle and notices they are all the same.

The proof by appeal to trigonometry will not be accessible to middle school students as they will not yet have learned any trigonometry, but high school students, if presented with the proof late enough in their trigonometry course might find it enlightening. At least it will exemplify the importance of some basic trigonometric laws by their applicability outside trigonometry.

The laws in question are a special case of the Pythagorean Theorem, a generalisation of the Pythagorean Theorem with error term that holds for non-right angles, and the Addition Formula for the Cosine. The special case of the Pythagorean Theorem is the familiar

$$\sin^2\theta + \cos^2\theta = 1,$$

from which we conclude

$$\sin\theta = \pm\sqrt{1 - \cos^2\theta}$$
$$= \sqrt{1 - \cos^2\theta}, \quad \text{if } 0 < \theta < 180°,$$

the inequality being the case if θ is one of the angles of a triangle.

The generalisation of the Pythagorean Theorem is called the Law of Cosines. To state it, recall the common notational convention that in a triangle ABC the side opposite a given vertex is denoted by the corresponding lower case Roman letter, thus a, b, c, respectively, and the angle at a given vertex is denoted by the corresponding lower case Greek letter, thus α, β, γ, respectively. With this, the *Law of Cosines* asserts

$$c^2 = a^2 + b^2 - 2ab\cos\gamma$$
$$b^2 = a^2 + c^2 - 2ac\cos\beta$$
$$a^2 = b^2 + c^2 - 2bc\cos\alpha.$$

Finally, the *Addition Formula for Cosines* reads

$$\cos(\theta + \vartheta) = \cos\theta\cos\vartheta - \sin\theta\sin\vartheta.$$

There are, of course, also the *Addition Formula for Sines*,

$$\sin(\theta + \vartheta) = \sin\theta\cos\vartheta + \sin\vartheta\cos\theta,$$

and a *Law of Sines* for a triangle ABC:

$$\frac{\sin\alpha}{a} = \frac{\sin\beta}{b} = \frac{\sin\gamma}{c}.$$

We shall not need these here, although I note that we can eliminate a little ugly algebra in our proof if we opt to use the Law of Sines in dealing with the sines later.

The proof is just a calculation of PQ, QR, and RP (See FIGURE 1.) and noting we get the same result in all three cases. The cases are symmetric, so I shall only present the details for the calculation of PQ. We begin by thinking

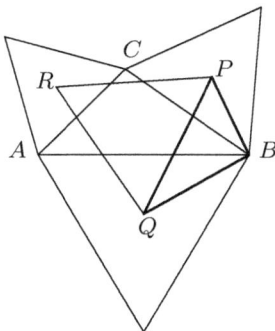

FIGURE 3.

about the triangle PBQ obtained in FIGURE 1 by filling in the lines PB and BQ as in FIGURE 3. By the Law of Cosines, we can express PQ in terms of PB, BQ, and $\cos(\angle PBQ)$:

$$(PQ)^2 = (PB)^2 + (BQ)^2 - 2 \cdot PB \cdot BQ \cdot \cos(\angle PBQ). \qquad (1)$$

The fact is that, because we are dealing with the centres of equilateral triangles, the quantities on the right can be calculated fairly easily.

To calculate PB, consider the equilateral triangle on side BC. The point P lies on the lines connecting the vertices to the midpoints of the opposite sides. In an equilateral triangle these lines are the perpendicular bisectors of the sides and, moreover, the bisectors of the angles of the triangle. Referring back to the original triangle ABC, BC is just a, whence the bisected segments have length $a/2$. Now with the bisector DE drawn in and labelled, we have FIGURE 4. The

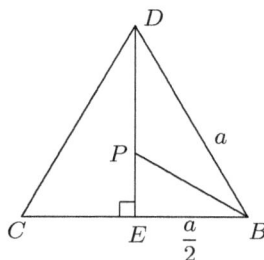

FIGURE 4.

length DE is quickly determined by the Pythagorean Theorem:

$$DE = \sqrt{a^2 - \left(\frac{a}{2}\right)^2} = \sqrt{\frac{4a^2 - a^2}{4}} = \frac{\sqrt{3}a}{2}.$$

Now the centre of a triangle lies on each bisector at a distance 2/3 the length of the bisector from the vertex. Moreover, in an equilateral triangle these bisectors have the same length. Thus,

$$PB = PD = \frac{2}{3} \cdot \frac{\sqrt{3}}{2} a = \frac{a}{\sqrt{3}}.$$

In the present case, however, we don't need to recall this fact: Triangle BPE is similar to DBE, whence

$$\frac{BP}{DB} = \frac{BE}{DE},$$

i.e.,

$$BP = DB \cdot \frac{BE}{DE} = a \cdot \frac{a/2}{(\sqrt{3}/2)a} = \frac{a}{\sqrt{3}}.$$

Likewise, $BQ = c/\sqrt{3}$.

Referring back to FIGURE 3 and formula (1), we have

$$(PQ)^2 = \frac{a^2}{3} + \frac{c^2}{3} - 2 \cdot \frac{a}{\sqrt{3}} \cdot \frac{c}{\sqrt{3}} \cdot \cos(\angle PBQ)$$
$$= \frac{a^2}{3} + \frac{c^2}{3} - \frac{2ac}{3} \cos(\angle PBQ). \tag{2}$$

Now

$$\angle PBQ = \angle PBC + \angle CBA + \angle ABQ$$
$$= 30° + \beta + 30° = \beta + 60°,$$

since PB bisects the 60° angle $\angle DBE$ and BQ likewise bisects a 60° angle. The Addition Formula yields

$$\cos(\angle PBQ) = \cos(\beta + 60)$$
$$= \cos\beta\cos 60 - \sin\beta\sin 60$$
$$= \frac{1}{2}\cos\beta - \frac{\sqrt{3}}{2}\sin\beta.$$

Plugging this into (2), we have

$$(PQ)^2 = \frac{a^2 + c^2}{3} - \frac{2ac}{3} \cdot \frac{1}{2}\cos\beta + \frac{2ac}{3} \cdot \frac{\sqrt{3}}{2}\sin\beta$$
$$= \frac{a^2 + c^2}{3} - \frac{ac}{3}\cos\beta + \frac{ac}{\sqrt{3}}\sin\beta. \tag{3}$$

Now formula (3) doesn't look all that promising. But we can eliminate $\cos\beta$ and $\sin\beta$ by going back to the original triangle ABC. First, we calculate $\cos\beta$ by applying the Law of Cosines to that triangle:

$$b^2 = a^2 + c^2 - 2ac\cos\beta,$$

whence

$$\cos \beta = \frac{b^2 - a^2 - c^2}{-2ac}.$$ (4)

Plugging this back into (3) yields

$$(PQ)^2 = \frac{a^2 + c^2}{3} - \frac{ac}{3} \cdot \frac{b^2 - a^2 - c^2}{-2ac} + \frac{ac}{\sqrt{3}} \sin \beta$$

$$= \frac{a^2 + c^2}{3} + \frac{b^2 - a^2 - c^2}{6} + \frac{ac}{\sqrt{3}} \sin \beta$$

$$= \frac{a^2 + b^2 + c^2}{6} + \frac{ac}{\sqrt{3}} \sin \beta$$ (5)

In like manner one obtains

$$(QR)^2 = \frac{a^2 + b^2 + c^2}{6} + \frac{bc}{\sqrt{3}} \sin \alpha$$ (6)

$$(RP)^2 = \frac{a^2 + b^2 + c^2}{6} + \frac{ab}{\sqrt{3}} \sin \gamma.$$ (7)

Now (5), (6), and (7) do not immediately appear to be equal. However, we can still express the sines in terms of the sides of triangle ABC:

$$\sin \beta = \sqrt{1 - \cos^2 \beta}$$

$$= \sqrt{1 - \left(\frac{b^2 - a^2 - c^2}{-2ac} \right)^2}$$

$$= \sqrt{\frac{4a^2c^2}{4a^2c^2} - \frac{b^4 + a^4 + c^4 - 2a^2b^2 - 2b^2c^2 + 2a^2c^2}{4a^2c^2}}$$

$$= \frac{\sqrt{2a^2b^2 + 2b^2c^2 + 2a^2c^2 - a^4 - b^4 - c^4}}{2ac},$$

whence

$$(PQ)^2 = \frac{a^2 + b^2 + c^2}{6} + \frac{1}{2\sqrt{3}} \sqrt{2a^2b^2 + 2b^2c^2 + 2a^2c^2 - a^4 - b^4 - c^4}$$

and the identical expressions occur when one replaces $\sin \alpha$, $\sin \gamma$ in (6) and (7), respectively.

This last bit of calculation is particularly ugly, and one can make the proof a bit more elegant by appeal to the Law of Sines,

$$\frac{ac}{\sqrt{3}} \sin \beta = \frac{ac}{\sqrt{3}} \cdot \frac{b \sin \alpha}{a} = \frac{bc}{\sqrt{3}} \sin \alpha,$$

whence (5) and (6) are equal, while

$$\frac{bc}{\sqrt{3}} \sin \alpha = \frac{bc}{\sqrt{3}} \cdot \frac{a \sin \gamma}{c} = \frac{ab}{\sqrt{3}} \sin \gamma,$$

and (6), (7) are equal.

Even with this more elegant replacement for the last step, the proof, like any computational one, is not particularly appealing or revealing. The equality of the three sides of Napoleon's triangle comes off looking accidental. Like any calculation, it is not easy to follow step-by-step. But, if one attempts to carry it out oneself, the computational steps are forced upon one and one merely needs FIGURE 3 and the instructions to find PQ in terms of PB, BQ and $\angle PBQ$ to get started. I consider the proof eminently reproducible, albeit not by those students who rely on memorisation. [Indeed, this is how I arrived at the present proof: Some years ago Gerard Theodore sent a short trigonometric proof of Napoleon's Theorem to the editors of *Philamath*, the newsletter of the Mathematical Study Unit of the American Topical Association, to which I belonged at the time. Monty J. Strauss, then president of the unit, asked me to embed the proof in a little exposition for inclusion in the newsletter, which I did[1]. The proof presented here is a slightly more detailed reproduction of Theodore's proof carried out without consulting my earlier write-up, which consultation, of course, would have been cheating.]

If you find the computational proof unsatisfying, look up a synthetic proof online. Compare the proof with the one given here. Which one starts with a more obvious strategy? Once underway, which proof leads you more directly to the final outcome? (I.e., do the successive steps stare you in the face, or are they all inspired?) Does the synthetic proof also appear serendipitous, or does it give you a real feel for why Napoleon's Theorem is true?

3 Project. Napoleon is not the only nonmathematician to have some mathematical connexion. I've listed a few at the beginning of this essay. Below is a list of some more. In each case, determine, where possible, i. the pursuit at which the individual is most famous, ii. the one in which he/she made his/her living, and iii. what the mathematical connexion was.

Adolf Anderssen	Tiburcio Carias Andino
John Dee	Charles Lutwidge Dodgson
Max Euwe	Benjamin Franklin
John Charles Fremont	Frederick II of Hohenstaufen
Alberto Fujimori	Gerbert of Aurillac
Thomas Harriot	Omar Khayyam
Samuel Pierpont Langley	Emanuel Lasker
Philipp Melanchthon	Story Musgrave
Dadabhai Naoroji	Florence Nightingale
Ronald Ross	Otto Yulievich Schmidt
Quintino Sella	Henrietta Szold
William Henry Fox Talbot	Paul Valery
Jan de Witt	Christopher Wren

4 Project. Many early mathematicians were involved in other sciences, e.g. astronomy. Some were also physicians (ibn Sina, Cardano) and some were artists.

---

[1] Craig Smoryński, "Napoleon's Theorem", *Philamath* 19, no. 3 (1998), pp. 3 - 4.

Geometry and perspective are obvious points of contact between mathematics and art. Are there others? Which famous artists can lay claim to the title of mathematician? Which mathematicians had their portraits painted by famous artists?

3

Some Trigonometry

The proof of Napoleon's Theorem given earlier depends on some trigonometry, a subject that originated with the Greeks and was further developed and transformed by the Arabs and Indians before being brought into final form by the Europeans. The specific results we used in proving Napoleon's Theorem are worth looking into.

The Law of Cosines is a variant of the Pythagorean Theorem, which latter result the Greeks probably originally proved by a simple dissection argument as was done by the Chinese and the Indians. Dissection arguments are intuitive and direct and thus are most likely the first proofs one would think of. For example, that the area of a triangle is half the base times the height is quickly verified for acute triangles by completing the rectangle as in FIGURE 1. The

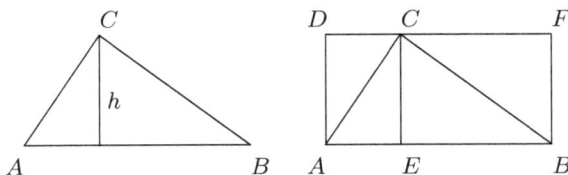

FIGURE 1.

congruences of the pairs AEC, CDA and EBC, FCB of triangles yields the equality of the areas in each pair:

$$\text{area } AEC = \text{area } CDA, \quad \text{area } EBC = \text{area } FCB.$$

But

$$\text{area } ABC = \text{area } AEC + \text{area } EBC$$
$$= \text{area } CDA + \text{area } FCB,$$

whence

$$\text{area } ABFD = \text{area } ABC + \text{area } BCF + \text{area } CDA$$
$$= \text{area } ABC + \text{area } ABC = 2 \cdot \text{area } ABC,$$

and we see that the area of the triangle is half that of the rectangle.

For an obtuse angle the argument is a bit tricker. One observes that the triangle is the difference of two right triangles as in FIGURE 2.

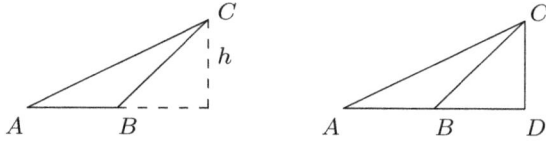

FIGURE 2.

One has

$$\text{area } ABC = \text{area } ADC - \text{area } BDC$$
$$= \frac{1}{2} \cdot AD \cdot h - \frac{1}{2} \cdot BD \cdot h$$
$$= \frac{1}{2}(AD - BD)h = \frac{1}{2} \cdot AB \cdot h.$$

Alternatively, one can generalise the treatment of the acute case by dealing with parallelograms as in FIGURE 3. This, of course, requires one to know that

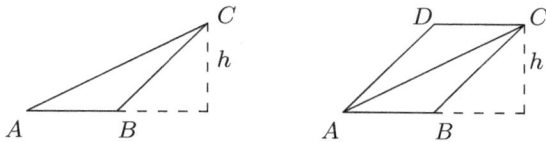

FIGURE 3.

the area of a parallelogram equals the base times the height, i.e., it equals the area of the rectangle on the given base whose other side equals the height. This latter equality can be established by a dissection argument. In the simple case, the top of the parallelogram overlaps with the top of the rectangle on the same base with the same height, as in FIGURE 4. In this case, one notices that the triangles DFA and CEB are congruent. For,

$$CD = AB = FE,$$

whence

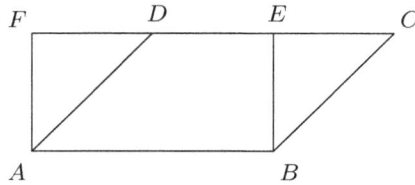

FIGURE 4.

$$FD = FE - DE = DC - DE = EC;$$
$$DA = CB;$$

and

$$\angle DFA = \angle CEB \text{ are right angles.}$$

Thus triangles DFA and CEB have the same area and

$$\begin{aligned} \text{area } ABCD &= \text{area } CEB + \text{area } EDAB \\ &= \text{area } DFA + \text{area } EDAB \\ &= \text{area } ABEF. \end{aligned}$$

The proof again is trickier when there is no overlap. One could observe that the parallelogram is the difference of two rectangles as in FIGURE 5. Thus

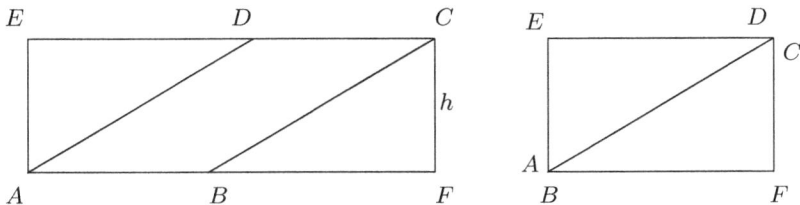

FIGURE 5.

$$\text{area } ABCD = AF \cdot CF - BF \cdot CF = AB \cdot h.$$

Alternatively, one could iterate the argument given in the overlapping case. Consider FIGURE 6. Here we draw in line BD and copy triangle CDB so that CB lies atop DA. Again DGA and CDB are congruent and the argument given in the overlapping case yields the equality of the areas of the parallelograms $ABCD$ and $ABDG$. By what was proven in the overlapping case, $ABDG$ has the same area as the rectangle $ABEF$, whence the area of $ABCD$ equals that of $ABEF$.

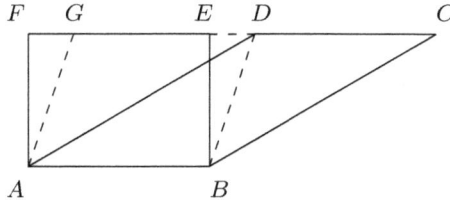

FIGURE 6.

Of course, there is no guarantee that the segment DG overlaps with EF and in general one might have to repeat the process several times, producing a chain of parallelograms of equal area, starting with $ABCD$ and ending with $ABEF$. Such a proof requires the *Axiom of Eudoxus*[1], better known as the *Archimedean Axiom*, asserting that given two unequal magnitudes, some multiple of the smaller exceeds the greater. In the present case, the supposedly larger magnitude would be the distance DE between the tops CD and EF of the parallelogram and the rectangle, respectively, and the smaller magnitude would be CD.

In *The Elements*, Euclid does not offer this proof. He postpones the use of the Archimedean Axiom to Book V, but proves the theorems about the areas of parallelograms and triangles in Book I. Oddly enough he doesn't use the subtraction argument cited above either. Instead he offers a third proof. He assumes as given two parallelograms on the same base and of equal height, or, as he puts it, with common base and "in the same parallels". There are two cases, depicted in FIGURE 7, according as there is or is not any overlap

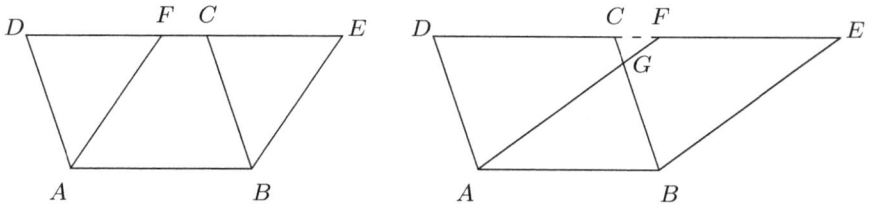

FIGURE 7.

of the tops of the parallelograms. Euclid treats only the more difficult of the two cases, namely that pictured on the right where there is no overlap. Using properties he has established for parallelograms, he concludes CD and EF to be equal, DA and CB to be equal, and $\angle FDA$ to equal $\angle ECB$. Adding FC separately to CD and EF, he concludes DF and EC to be equal. It follows that the triangles AFD and BEC are congruent and have equal area. Removing the

[1] Due to Eudoxus of Cnidus (*c.* 400 B.C. - *c.* 347 B.C.)

common part, triangle FCG, from each of these yields two trapeziums[2] $AGCD$ and $BEFG$ of equal area. Adding the triangle ABG to each of these yields the parallelograms $ABCD$ and $ABEF$, respectively, showing them to have equal area.

The easy case of the left half of FIGURE 7 can be proven similarly, but without the subtraction of any overlapping triangles. Alternatively, one can reduce this case to the previous one by comparing the areas of the two parallelograms to a third one whose top has no overlap with either.

Before moving on to discuss Euclid's proof of the Pythagorean Theorem, let us pause to consider the humble trapezium. One of the worst things one can do pædagogically is to teach students to memorise formulæ. Some formulæ have to be remembered, like those for the area of a rectangle and the area of a circle. The formula for the area of a trapezium,

$$A = \frac{1}{2}(B + b)h,$$

where

h is the altitude

B is the larger base

b is the smaller base,

is not one of the formulæ that should be memorised. It may stick in the student's brain through most of a course, but unless the student is taught why the formula holds, or practises finding the areas of numerous trapeziums, it is not likely to be retained after that. I am inclined to think, however, that a student who has practised dissecting a few parallelograms and triangles will have no problem when confronted with a trapezium, say that of FIGURE 8, below.

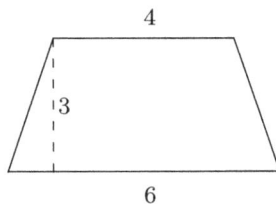

FIGURE 8.

Getting back on track, I note that Euclid proved the Pythagorean Theorem by a sort of dissection argument. He did not use any of the simple arguments

---

[2] This is one of those instances where British and American English disagree. In both variants of the language, a trapezium is a quadrilateral with one pair of sides parallel and the other not. In American usage this is more commonly called a trapezoid, while in Britain no two sides of a trapezoid are parallel. The most popular English translation of Euclid is Thomas Little Heath's (1861 - 1940) British one, in which the quadrilaterals in question are called trapeziums.

found in India or China, but rather used his knowledge of the areas of triangles and parallelograms to show the areas of the smaller squares $ACHI$ and $BFGC$ to equal those of the rectangles $ADJK$ and $KJEB$, respectively. (See FIGURE 9.)

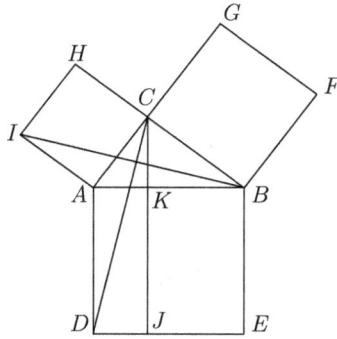

FIGURE 9.

Euclid establishes the equality of $ADJK$ and $ACHI$ by noting the following:

    i. triangle ADC has base AD and height AK, whence equals half the rectangle $ADJK$ in area;

    ii. triangle IAB has base IA and height AC, whence equals half the rectangle $ACHI$ in area;

    iii. triangles IAB and CAD are congruent sharing two pairs of equal sides (IA and CA, AB and AD) and the angle between them ($\angle IAB$ and $\angle CAD$ each equals $\angle CAB$ plus a right angle) and thus have equal area.

The argument for the equality of $KJEB$ and $BFGC$ is analogous.

After establishing the Pythagorean Theorem in this way as the 47th proposition of the first book of *The Elements*, Euclid finishes the book with the converse proposition: if ABC is a triangle in which the square on one side equals the sum of the squares on the remaining two sides, then the triangle is a right triangle. The result is given a clever, but not particularly revealing proof: If $AB^2 + BC^2 = AC^2$, construct a right triangle with legs equal to AB and BC. By the Pythagorean Theorem, the hypotenuse equals AC, whence the two triangles are congruent and ABC is indeed a right triangle. Euclid would do better in Book II.

Book II is Euclid's now controversial development of geometric algebra, so called because the mathematics is the algebra of quadratic equations. The book has the ultimate goal of proving Propositions II-14 asserting that any polygonal figure can be made equal in area to a square. The two penultimate results form the following pair of results complementing the Pythagorean Theorem.

1 Theorem (Propositions II-12). *In obtuse-angled triangles the square on the side subtending the obtuse angle is greater than the squares on the sides containing the obtuse angle by twice the rectangle contained by one of the sides about the obtuse angle, namely that on which the perpendicular falls, and the straight line cut off outside by the perpendicular towards the obtuse angle.*

2 Theorem (Proposition II-13). *In acute-angled triangles the square on the side subtending the acute angle is less than the squares on the sides containing the acute angle by twice the rectangle contained by one of the sides about the acute angle, namely that on which the perpendicular falls, and the straight line cut off within by the perpendicular towards the acute angle.*

Modulo relabelling for the sake of uniformity, the Euclidean diagrams illustrating these propositions are reproduced in FIGURE 10. Proposition II-12

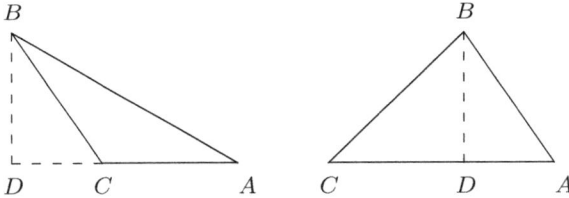

FIGURE 10.

asserts

$$AB^2 = BC^2 + CA^2 + 2 \cdot AC \cdot DC \qquad (1)$$

in the left half of the diagram; while II-13 says

$$AB^2 = BC^2 + CA^2 - 2 \cdot AC \cdot DC \qquad (2)$$

in the right half. Writing a, b, c for the sides opposite the vertices A, B, C, respectively, these read

$$\left. \begin{array}{l} c^2 = a^2 + b^2 + 2 \cdot b \cdot DC \\ c^2 = a^2 + b^2 - 2 \cdot b \cdot DC, \end{array} \right\} \qquad (3)$$

respectively. Now these are just the obtuse and acute cases, respectively, of the Law of Cosines. For, if γ is the angle $\angle BCA$,

$$\cos \gamma = \begin{cases} -\frac{DC}{BC}, & \gamma \text{ obtuse} \\ \frac{DC}{BC}, & \gamma \text{ acute,} \end{cases}$$

whence

$$DC = \begin{cases} -BC \cos \gamma = -a \cos \gamma, & \gamma \text{ obtuse} \\ BC \cos \gamma = a \cos \gamma, & \gamma \text{ acute,} \end{cases}$$

and (3) uniformly reads

$$c^2 = a^2 + b^2 - 2ab\cos\gamma. \tag{4}$$

The full Law of Cosines, extending (4) to the case in which γ is a right angle, follows from the Pythagorean Theorem and the fact that $\cos\gamma = 0$ for right angles γ.

The proofs of Propositions II-12 and II-13, and therewith of the Law of Cosines, are quite easy. Applying the Pythagorean Theorem to triangle ABD in the obtuse case yields

$$\begin{aligned}
AB^2 &= BD^2 + AD^2 \\
&= (BC^2 - DC^2) + (DC + AC)^2 \\
&= BC^2 - DC^2 + DC^2 + 2 \cdot DC \cdot AC + AC^2 \\
&= BC^2 + AC^2 + 2 \cdot DC \cdot AC,
\end{aligned}$$

which is just (1). In the acute case we have

$$\begin{aligned}
AB^2 &= BD^2 + DA^2 \\
&= (BC^2 - CD^2) + (AC - CD)^2 \\
&= BC^2 - CD^2 + AC^2 - 2 \cdot AC \cdot CD + CD^2 \\
&= BC^2 + AC^2 - 2 \cdot CD \cdot AC,
\end{aligned}$$

which is (2).

We didn't really need the Law of Sines in discussing Napoleon's Theorem, but it is traditional to discuss it together with the Law of Cosines, so we might as well do so here. Not being a variant of the ubiquitous Pythagorean Theorem, it is less memorable. Its proof, however, is much easier. Consider a triangle ABC with angles α, β, γ at A, B, C, respectively, and sides a, b, c opposite them, and drop an altitude CD from vertex C to the opposite side as in FIGURE 11. In

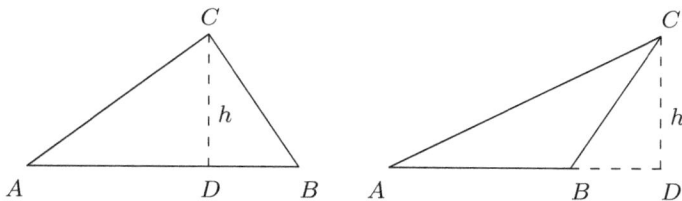

FIGURE 11.

keeping with our discussion of areas, we calculate the area of the triangle

$$\text{Area} = \frac{1}{2} \cdot h \cdot AB = \frac{1}{2}(b\sin\alpha)c = \frac{1}{2}abc\frac{\sin\alpha}{a}.$$

The same calculation, seen from the other angles, yields

$$\text{Area} = \frac{1}{2}abc\frac{\sin\beta}{b} = \frac{1}{2}abc\frac{\sin\gamma}{c}.$$

Cancelling the common factor from the three expressions for the area yields

$$\frac{\sin\alpha}{a} = \frac{\sin\beta}{b} = \frac{\sin\gamma}{c},$$

which is the Law of Sines.

Actually, the key to the proof is the calculation not of the area but of the altitude h:

$$h = AC\sin\alpha = b\sin\alpha$$
$$h = BC\sin\beta = a\sin\beta,$$

whence $a\sin\beta = b\sin\alpha$ and

$$\frac{\sin\alpha}{a} = \frac{\sin\beta}{b}.$$

Similarly,

$$\frac{\sin\beta}{b} = \frac{\sin\gamma}{c},$$

and the Law is established.

Historically, the declaration that the two propositions II-12 and II-13 of Euclid, together with the Pythagorean Theorem, comprise the Law of Cosines is an anachronism. Sines and cosines were only first introduced some centuries later in India. Greek trigonometry concerned itself with the *chord*, the chord of an arc $\overset{\frown}{AB}$ of a circle being the line segment AB joining the endpoints A, B of the arc. Moreover, Greek trigonometry originated with Hipparchus of Bithynia (*fl.* after 127 B.C.) over a century after Euclid's time. Fundamental results on which Greek trigonometry rests already appear, however, in Book III of *The Elements*.

The third book of *The Elements* deals with circles and lines—diameters, tangents, chords, and the sides of inscribed and inscribing figures. Of particular importance here is the following.

3 Theorem (Proposition III-20). *In a circle the angle at the centre is double of the angle at the circumference, when the angles have the same circumference as base.*

What this means is best explained by a picture—actually by three pictures as in FIGURE 12. Euclid's proposition asserts that, in each case, $\angle BCA$ is half $\angle BOA$. His own illustration is a cluttered diagram depicting all three cases and he obscures the proof by going straight for the case illustrated by the second diagram, proving the case given in the first diagram as he proceeds—but not mentioning this. The key to the entire proof, however, is the first, simple case.

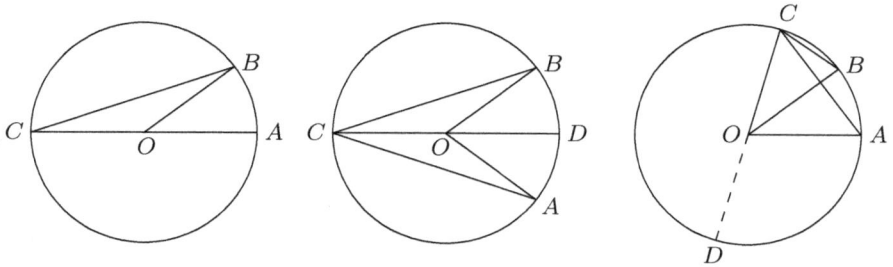

FIGURE 12.

Assume A, B, C are as in the first circle of FIGURE 12. Because AOC is a straight line, $\angle COB$ and $\angle BOA$ add up to two right angles. But the angles of a triangle add up to two right angles, whence

$$\angle BCO + \angle COB + \angle OBC = \angle COB + \angle BOA,$$

i.e.,

$$\angle BCO + \angle OBC = \angle BOA.$$

However, triangle COB is isosceles with the radii CO and OB being equal sides. Thus $\angle BCO = \angle OBC$ and

$$\angle BOA = 2 \cdot \angle BCO = 2 \cdot \angle BCA,$$

as was to be proved.

In the case illustrated by the centre diagram, apply the case just proven:

$$\angle BCA = \angle BCD + \angle DCA$$
$$= \frac{1}{2} \cdot \angle BOD + \frac{1}{2} \cdot \angle DOA = \frac{1}{2}(\angle BOD + \angle DOA)$$
$$= \frac{1}{2} \cdot \angle BOA.$$

And do the same for the third configuration:

$$\angle BCA = \angle BCD - \angle DCA$$
$$= \frac{1}{2} \cdot \angle BOD - \frac{1}{2} \cdot \angle DOA = \frac{1}{2}(\angle BOD - \angle DOA)$$
$$= \frac{1}{2} \cdot \angle BOA.$$

By 20th and 21st century tastes, the functional dependence of the length of a chord on the angle of an arc (given a fixed radius) is first on our minds. Euclid establishes this over a series of propositions. His very next proposition, for instance, states that "the angles in the same segment are equal to one another", i.e., if we look at FIGURE 13, $\angle ACB = \angle ADB$. The reason, of

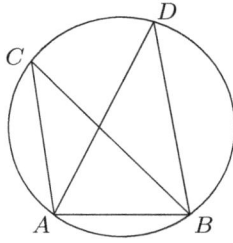

FIGURE 13.

course, is that each angle is $\frac{1}{2}\angle AOB$ obtained when we add the centre O of the circle to the diagram along with the radii OA, OB.

One immediate corollary to Theorem 3 that Euclid takes great pains to prove, probably because he does not consider the 180° angle to be an angle, is that, if a triangle is inscribed in a circle in such a way that one of its sides is a diameter, then the triangle is a right triangle. Yet this is immediate, for if ABC is a triangle with AB constituting the diameter of a circle, as in FIGURE 14, then $\angle ACB$ is half of angle $\angle AOB$ which, in Euclidean parlance, is two

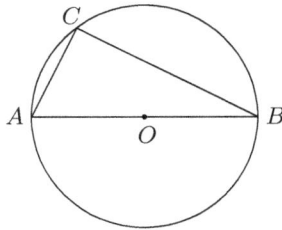

FIGURE 14.

right angles.

We will use this result below.

Getting back to trigonometry, Hipparchus, in his work on astronomy, found occasion to compute a table of chords. In this undertaking, he followed the Babylonians in dividing the circle into 360 equal parts, which, of course, we now call degrees. The numerical value of the chord, i.e., its length, depends on the radius of the circle chosen. The astronomers we shall mention here—Hipparchus, Ptolemy (*c.* 100 - 178 A.D.), Copernicus (1473 - 1543)—all used different values for the radius. Today the standard is a circle of radius 1, so we shall henceforth assume this as our radius.

Hipparchus, whose work does not survive, calculated a table of chords by determining directly the chords of a few specific angles and filling in as much as he could by calculating the chords of the half-angles of angles already in his table. It is assumed he used the correct formula here, as it is not that difficult to derive. (Indeed, we did so in discussing the value of π.)

He was also able to fill in some values by using the supplementary angle formula

$$\text{chord}(180 - \alpha) = \sqrt{4 - \text{chord}^2\,\alpha} \tag{5}$$

which is an instance of the Pythagorean Theorem. If $\angle BOC$ is complementary to $\angle AOB$, the figure ABC in FIGURE 15 is a right triangle, whence

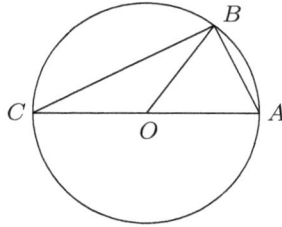

FIGURE 15.

$$BC^2 = AC^2 - AB^2 = 4 - AB^2$$

if we assume a radius of 1, and $BC = \sqrt{4 - AB^2}$, i.e., (5).

The greatest work of astronomy of the ancient world, and coincidentally of trigonometry, is *The Mathematical Composition* of Claudius Ptolemy. The book came to be called *The Great Astronomer* to distinguish it from a shorter text called *The Little Astronomer*. The Arabs, some centuries later, took to calling it "The Greatest" in a bilingual manner, combining the Arabic article *al* with the Greek *megiste*. The resulting name, *Almagest*, is that by which the book is known today.[3] Book I of this work begins with 9 sections of astronomical assumptions before coming to the mathematics of sections 10 and 11. Section 10, titled "On the size of chords in a circle", contains geometric demonstrations of propositions on chords, which are used to compute the chords of various angles. The results of these computations are gathered together in section 11 as a table of chords calculated to 4 sexigesimal places for arcs from $\frac{1}{2}$ to 180 degrees in half-degree increments. To carry out this construction, he calculated directly the chords of a few specific angles and then obtained the rest by means of geometrically established rules for calculating the sums, differences, and the halves and complements of angles whose chords are already known.[4]

Ptolemy begins with the following proposition.

---

[3] Sometimes it is called *The Almagest*, redundantly tacking on an extra article and, incidentally, making the title trilingual.

[4] To be honest, he does this down to $1\frac{1}{2}$ degrees and relies on an interpolation argument to get down to half a degree.

4 Theorem (Ptolemy's Theorem). *Let the quadrilateral $ABCD$ be inscribed in a circle as in* FIGURE 16, *below. Then:* $AC{\cdot}BD = AB{\cdot}DC + AD{\cdot}BC$.

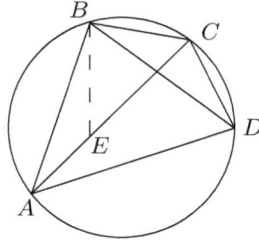

FIGURE 16.

Actually, although he would clearly be multiplying numbers, he expressed the equation geometrically:

$$\text{rect. } AC, BD = \text{rect. } AB, DC + \text{rect. } AD, BC.$$

Copernicus, in copying the proof in *De revolutionibus orbium coelestium* (1543), still expressed it geometrically as follows:

If a quadrilateral is inscribed in a circle, the rectangle comprehended by the diagonals is equal to the two rectangles which are comprehended by the two pairs of opposite sides.

Proof of Theorem 4. The proofs offered by Ptolemy and Copernicus are the same. Drop a line BE from B to diagonal AC in such a way that $\angle ABE = \angle DBC$. Now

$$\angle ABD = \angle ABE + \angle EBD = \angle DBC + \angle EBD = \angle EBD + \angle DBC = \angle EBC.$$

But also $\angle BDA = \angle BCA$ by Theorem 3 since the two angles share the chord AB. Thus triangles ABD and EBC are similar.

One now compares ratios:

$$\frac{BC}{CE} = \frac{BD}{AD},$$

whence

$$AD \cdot BC = BD \cdot CE. \qquad (6)$$

The next step is to show that triangles ABE and DBC are similar. The equality of $\angle ABE$ and $\angle DBC$ was imposed in constructing BE. But $\angle EAB = \angle CAB$ which shares the chord CB with $\angle CDB$. Thus $\angle CAB = \angle CDB$ and $\angle EAB = \angle CDB$. Hence the triangles ABE and DBC are similar and again we have

$$\frac{AB}{AE} = \frac{DB}{DC},$$

whence

$$AB \cdot DC = AE \cdot DB. \tag{7}$$

Combining (6) and (7) we have

$$AB \cdot DC + AD \cdot BC = AE \cdot DB + BD \cdot CE = AC \cdot DB,$$

since $AC = AE + CE$. □

In time-honoured fashion, Ptolemy's proof is incomplete. There are two other diagrams to consider—that in which BE falls inside the angle $\angle CBD$ instead of inside $\angle ABD$, and that in which BE falls on the line BD. I shall follow Ptolemy's lead and leave these to my readers to ruminate over.

The reader who has not lost sight of our goal of looking into the results upon which Napoleon's Theorem rests may be wondering if I have done so. Well, the second result, after the Law of Cosines, that we used was the Addition Formula for Cosines. And Theorem 4 provides us with an Addition Formula for Chords if we choose our quadrilateral carefully.

Let $\widehat{AB}$ and $\widehat{BC}$ be arcs of the unit circle centred at O, each arc less than $180°$, and let D lie on the circumference in the straight line with BO, so that BOD is a diameter of the circle, as in FIGURE 17 and draw in the chords

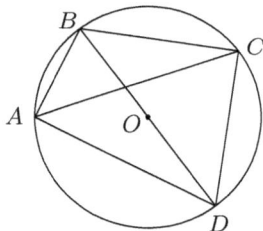

FIGURE 17.

AB, BC, CD, DA and AC. By Ptolemy's Theorem,

$$AC \cdot BD = AB \cdot CD + BC \cdot AD,$$

i.e.,

$$\text{chord}(\widehat{AB} + \widehat{BC}) \cdot 2 = \text{chord}(\widehat{AB}) \cdot \text{chord}(\widehat{CD}) + \text{chord}(\widehat{BC}) \cdot \text{chord}(\widehat{AD}),$$

i.e.,

$$\text{chord}(\alpha + \beta) = \frac{1}{2}\big(\text{chord}(\alpha)\,\text{chord}(180 - \beta) + \text{chord}(\beta)\,\text{chord}(180 - \alpha)\big), \tag{8}$$

where α, β are the central angles $\angle AOB, \angle BOC$, respectively.

Ptolemy does not immediately spell this out for us. Instead he applies his theorem to give a sort of Subtraction Formula. He draws FIGURE 18, below, and says

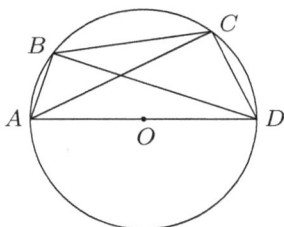

Now that this [i.e., Theorem 4] has been expounded, let there be the semicircle $ABCD$ on diameter AD, and from point A let there be drawn two straight lines AB, AC, and let the lengths of each of them be given...
I say that BC is also given.
For let BD and CD be joined. Then clearly they are also given because they subtend the supplements. Since, then, the quadrilateral $ABCD$ is inscribed in a circle, therefore

$$\text{rect. } AB, CD + \text{rect. } AD, BC = \text{rect. } AC, BD.$$

And rectangle AC, BD is given, and also rectangle AB, CD. Therefore the remaining rectangle AD, BC is also given. And AD is the diameter. Hence the straight line BC is also given. And it is now clear to us that, if two arcs are given and the two chords subtending them, then also the chord subtending the difference between the two arcs will be given.

I'm not sure how clear all this is. Geometrically, it is obvious that BC is "given" from AB and AC once one has drawn the picture. It is the arithmetic determination that is in question and which is rather loosely expressed. Copernicus is clearer in his treatment. First he offers a loose formal statement of the result as a corollary to Theorem 4

5 Theorem. *Hence if straight lines subtending unequal arcs in a semicircle are given, the chord subtending the arc whereby the greater exceeds the smaller is also given.*

The meaning of this, i.e., the sense in which BC is given once AC, AB are given, is clarified by the proof. From Theorem 4 we have

$$\text{rect. } AC, BD = \text{rect. } AB, CD + \text{rect. } AD, BC$$

and thus

$$\text{rect. } AD, BC = \text{rect. } AC, BD - \text{rect. } AB, CD.$$

And, "accordingly, in so far as the division may be carried out,

$$(AC \cdot BD - AB \cdot CD) \div AD = BC, \tag{9}$$

which was sought".

If we write α, β, γ for the angles determining arcs $\widehat{AB}, \widehat{BC}, \widehat{AC}$, respectively, we have $\beta = \gamma - \alpha$ and (9) reads

$$\text{chord}(\gamma - \alpha) = \frac{\text{chord}(\gamma)\,\text{chord}(180 - \alpha) - \text{chord}(\alpha)\,\text{chord}(180 - \gamma)}{2}. \tag{10}$$

Ptolemy and Copernicus both follow this result with the half-angle formula giving the chord of $\alpha/2$ in terms of chord α. Only then do Ptolemy and Copernicus show how to find $\text{chord}(\alpha + \beta)$ from chord α and chord β. Oddly enough, they do not derive (8). Instead they draw a new diagram (FIGURE 19) assuming

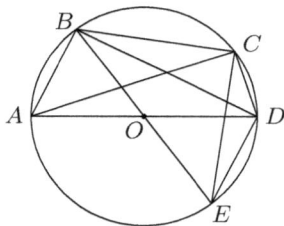

FIGURE 19.

$\widehat{AB} + \widehat{BC} < 180$. They now note essentially that

$$CE = \text{chord}(180 - \widehat{BC})$$
$$BD = \text{chord}(180 - \widehat{AB})$$
$$DE = AB,$$

This last equality follows from the congruence of the triangles AOB and DOE ($\angle AOB = \angle DOE$ and the sides AO, BO, DO, EO are all radii, whence equal). They now apply Ptolemy's Theorem to the quadrilateral $BCDE$ to conclude

$$BD \cdot CE = BC \cdot DE + CD \cdot BE,$$

i.e.,

$$\text{chord}(180 - \widehat{AB})\,\text{chord}(180 - \widehat{BC}) = BC \cdot AB + 2CD.$$

Writing α, β for the angles yielding arcs $\overset{\frown}{AB}, \overset{\frown}{BC}$, respectively, this reads

$$\mathrm{chord}(180 - \alpha)\,\mathrm{chord}(180 - \beta) = \mathrm{chord}(\beta)\,\mathrm{chord}(\alpha) + 2\,\mathrm{chord}(180 - (\alpha + \beta)),$$

i.e.,

$$\mathrm{chord}(180 - (\alpha + \beta)) = \frac{1}{2}\big(\,\mathrm{chord}(180 - \alpha)\,\mathrm{chord}(180 - \beta) - \mathrm{chord}(\alpha)\,\mathrm{chord}(\beta)\big),$$
$$(11)$$

rather a strange looking substitute for (8).

Our goal, of course, is to obtain the Addition Formulæ for the more familiar sine and cosine. With one major caveat, this is essentially a matter of translation. For, for small enough α, we have

$$\sin \alpha = \frac{1}{2}\,\mathrm{chord}(2\alpha)$$

or, put differently,

$$\mathrm{chord}\,\alpha = 2\sin(\alpha/2).$$

A glance at FIGURE 20 should suffice to convince the reader of the truth of

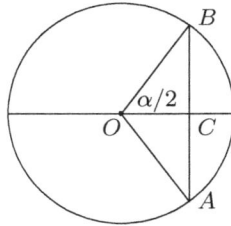

this. If α is the central angle $\angle AOB$ yielding AB as chord, and OC is the perpendicular bisector of AB, then $\angle BOC$ is half of $\angle BOA$, i.e., $\alpha/2$, and $\sin(\alpha/2)$ is $BC = (1/2)AB$.

Formula (8) yields

$$
\begin{aligned}
\sin(\alpha + \beta) &= \frac{1}{2}\,\mathrm{chord}(2\alpha + 2\beta) \\
&= \frac{1}{4}\big(\,\mathrm{chord}(2\alpha)\,\mathrm{chord}(180 - 2\beta) + \mathrm{chord}(2\beta)\,\mathrm{chord}(180 - 2\alpha)\big) \\
&= \frac{1}{4}\big(2\sin(\alpha)2\sin(90 - \beta) + 2\sin(\beta)2\sin(90 - \alpha)\big) \\
&= \sin \alpha \cos \beta + \sin \beta \cos \alpha, \quad\quad\quad\quad\quad\quad\quad\quad\quad\quad\quad (12)
\end{aligned}
$$

using the identity

$$\sin(90 - \gamma) = \cos \gamma, \text{ for } 0 \le \gamma \le 90.$$

Thus (8) yields the Addition Formula for Sines.

Formula (10) yields

$$
\begin{aligned}
\sin(\gamma - \alpha) &= \frac{1}{2} \operatorname{chord}(2\gamma - 2\alpha) \\
&= \frac{1}{4} \big(\operatorname{chord}(2\gamma) \operatorname{chord}(180 - 2\alpha) - \operatorname{chord}(2\alpha) \operatorname{chord}(180 - 2\gamma) \big) \\
&= \frac{1}{4} \big(2\sin(\gamma) 2 \sin(90 - \alpha) + 2 \sin(\alpha) 2 \sin(90 - \gamma) \big) \\
&= \sin \gamma \cos \alpha - \sin \alpha \cos \gamma,
\end{aligned}
\tag{13}
$$

the Subtraction Formula for Sines. And (11) yields

$$
\begin{aligned}
\cos(\alpha + \beta) &= \sin(90 - (\alpha + \beta)) = \frac{1}{2} \operatorname{chord}(180 - (2\alpha + 2\beta)) \\
&= \frac{1}{4} \big(\operatorname{chord}(180 - 2\alpha) \operatorname{chord}(180 - 2\beta) - \operatorname{chord}(2\alpha) \operatorname{chord}(2\beta) \big) \\
&= \frac{1}{4} \big(2 \sin(90 - \alpha) 2 \sin(90 - \beta) - 2 \sin \alpha \, 2 \sin \beta \big) \\
&= \cos \alpha \cos \beta - \sin \alpha \sin \beta,
\end{aligned}
\tag{14}
$$

the Addition Formula for Cosines.

The only thing missing is the Subtraction Formula for Cosines, which is easy enough:

$$
\begin{aligned}
\cos(\gamma - \alpha) &= \sin(90 - (\gamma - \alpha)) = \sin((90 - \gamma) + \alpha) \\
&= \sin(90 - \gamma) \cos \alpha + \cos(90 - \gamma) \sin \alpha, \text{ by (12)} \\
&= \cos \gamma \cos \alpha + \sin \gamma \sin \alpha.
\end{aligned}
\tag{15}
$$

The big caveat to all of this is that we've only established this for small angles. Ptolemy constructed tables of chords for angles between 0 and 180 degrees, thus his $\alpha, \beta, \alpha + \beta$ were all restricted to lying within this range. Likewise, Copernicus constructed tables of sines (labelled half the chord of the double arc) for angles between 0 and 90 degrees, thus his $\alpha, \beta, \alpha + \beta$ lay within this narrower range. Extending (12) - (15) beyond this range requires sines and cosines to be defined for negative angles (e.g., in identifying $\cos \alpha$ with $\sin(90 - \alpha)$ for $\alpha > 90$) and in assigning negative values to $\sin \alpha$ and $\cos \alpha$ for certain angles α. This sort of thing would have to come later. For Ptolemy and Copernicus chords were line segments or their (positive) lengths.

Thus we have not established the Addition Formula for Cosines in as strong a form as needed for the applications we made of it in discussing Napoleon's Theorem. We can get around this by drawing a new diagram (FIGURE 21) and considering the quadrilateral $BDCE$. We have

$$BC \cdot DE = BD \cdot CE + DC \cdot BE.$$

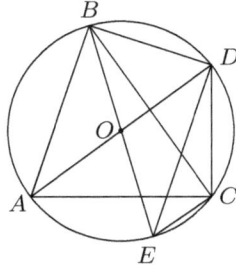

FIGURE 21.

Letting $\angle AOB = \alpha, \angle BOC = \beta$, we have

$$BC = \text{chord } \beta$$
$$DE = AB = \text{chord } \alpha$$
$$BD = \text{chord}(180 - \alpha)$$
$$CE = \text{chord}(180 - \beta)$$
$$DC = \text{chord}(\alpha + \beta - 180)$$
$$BE = 2$$

Thus

$$\text{chord}(180 - \alpha)\,\text{chord}(180 - \beta) + 2\,\text{chord}(\alpha + \beta - 180) = \text{chord}(\alpha)\,\text{chord}(\beta)$$

i.e.,

$$2\,\text{chord}(\alpha + \beta - 180) = \text{chord}\,\alpha\,\text{chord}\,\beta - \text{chord}(180 - \alpha)\,\text{chord}(180 - \beta). \quad (16)$$

Replacing α, β by $2\alpha, 2\beta$, respectively, this yields

$$\sin(\alpha + \beta - 90) = \sin \alpha \sin \beta - \cos \alpha \cos \beta. \quad (17)$$

But by FIGURE 22, it is clear that $\sin(\alpha + \beta - 90) = -\cos(\alpha + \beta)$, whence (17) yields

$$-\cos(\alpha + \beta) = -\cos \alpha \cos \beta + \sin \alpha \sin \beta,$$

i.e.

$$\cos(\alpha + \beta) = \cos \alpha \cos \beta - \sin \alpha \sin \beta$$

for $90 \leq \alpha + \beta \leq 180$.

Thinking about all the cases that have to be proven separately makes one appreciate higher theory. In Analysis, for example, one can *define* $\sin x$ and $\cos x$ by their infinite *power series* expansions (where x is given in *radians* instead of degrees):

$$\sin x = x - \frac{x^3}{3!} + \frac{x^5}{5!} - \frac{x^7}{7!} + \dots \quad (18)$$

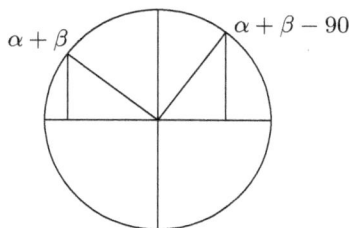

FIGURE 22.

$$\cos x = 1 - \frac{x^2}{2!} + \frac{x^4}{4!} - \frac{x^6}{6!} + \ldots \qquad (19)$$

and notice that their derivatives are

$$\frac{d \sin x}{dx} = \cos x, \quad \frac{d \cos x}{dx} = -\sin x.$$

Thus, if $f(x)$ is $\sin x$ or $\cos x$, the second derivative satisfies

$$f''(x) = -f(x), \qquad (20)$$

and any solution to (20) must be a linear combination of $\sin x$ and $\cos x$:

$$f(x) = a \sin x + b \cos x \qquad (21)$$

for some constants a, b. Now if we take $f(x) = \sin(x + \beta)$, the Chain Rule in Calculus tells us that

$$f'(x) = \cos(x + \beta), \quad f''(x) = -\sin(x + \beta),$$

i.e., f satisfies (20) and is thus of the form (21). Now

$$f'(x) = a \cos x - b \sin x \qquad (22)$$

and using (21) -(22) we have

$$\sin \beta = f(0) = a \sin 0 + b \cos 0 = a \cdot 0 + b \cdot 1 = b$$
$$\cos \beta = f'(0) = a \cos 0 - b \sin 0 = a \cdot 1 - b \cdot 0 = a.$$

Thus

$$\sin(x + \beta) = \cos \beta \sin x + \sin \beta \cos x,$$

which is the Addition Formula for Sines for all values of x, β. That for cosine and the Subtraction Formulæ can be proven similarly.

Of course, if one is not familiar with the Calculus, Differential Equations, and Linear Algebra, none of this is very meaningful. And one is stuck with the question: how do we know that (18) and (19) are the familiar trigonometric

functions? It is an amusing exercise in the higher Calculus to develop trigonometry from these definitions and ultimately prove that, for $\alpha < 2\pi$, the point
$(\cos\alpha, \sin\alpha)$ is indeed the point on the unit circle ending an arc of length α beginning at $(1,0)$. I refer the reader with sufficient background to Walter Rudin,
Principles of Mathematical Analysis, 2nd ed., McGraw-Hill Book Company,
New York, 1964, pp. 167 - 169 for the details.

This last approach uses, perhaps, too much high-powered machinery. One
might try simple analytic geometry, defining, for any angle α, the values
$\cos\alpha, \sin\alpha$ to be the x- and y-coordinates of the point on the unit circle lying on
the ray starting at the origin and making an angle α with the nonnegative real
axis—where one goes counter-clockwise from the x-axis to the ray for positive
α and clockwise for negative α. However, when one draws the picture (FIGURE
23), one realises that there will be a number of cases to consider according to

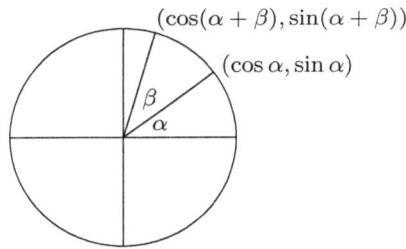

FIGURE 23.

which quadrants $\alpha, \beta, \alpha + \beta$ lie in.

I should also mention that it is possible to prove the Addition Formulæ
for Sines and Cosines geometrically without recourse to chords. One proof,
due to the Arabic mathematician Abū'l-Wafā, can be found in J.L. Berggren's
Episodes in the Mathematics of Medieval Islam[5].

---

[5] Springer-Verlag, New York, 1986, pp. 135 - 138.

Part II

Computation

4

The Pell Equation

Berlinghoff and Gouvêa note on page 28 of their text that in the 7th century Brahmagupta (598 - after 665) had considered the problem of finding integral solutions to the equation

$$92X^2 + 1 = Y^2. \tag{1}$$

Nowadays we would interchange X and Y and write

$$X^2 - 92Y^2 = 1. \tag{2}$$

This is a special case of a more general equation,

$$X^2 - dY^2 = 1, \tag{3}$$

where d is a positive integer which is not a perfect square. This equation in turn is a special case of the more general,

$$X^2 - dY^2 = c, \tag{4}$$

where d is again a positive integer that is not a perfect square and where c is an integer. Equation (3) is called the *Pell equation* and is of great interest. Equation (4) may also be called a Pell equation, but for $c \neq \pm 1, \pm 2$ it is generally only of auxiliary interest for use in solving (3). The Pell equation has a long history and is important in the history of mathematics.

The importance of the Pell equation is easy to see when one assumes x, y to be a solution and divides the equation by y^2:

$$\frac{x^2}{y^2} - d = \frac{1}{y^2}, \quad \text{i.e.,} \quad \left(\frac{x}{y}\right)^2 = d + \frac{1}{y^2}.$$

Thus, if y is large enough, x/y is a very good approximation to $\sqrt{d}$. Many of the early values recorded for square roots are of this form: fractions whose numerators and denominators form solutions to the corresponding Pell equations. This has led to the conjecture that some instances of the Pell equation were known and considered early on.

The more definite histories of the Pell equation begin with Archimedes. The 19th century poet and dramatist Gottfried Ephraim Lessing worked for a period as librarian at the library of Duke August in Wolfenbüttel, the same library at which Leibniz had earlier worked and in which the famous Wolfenbüttel manuscript of the old Faust *Volksbuch* is to be found. It was there that Lessing discovered a puzzle attributed to Archimedes. The puzzle, called the *Cattle Problem of Archimedes*, is a complicated word problem involving a number of variables and its solution ultimately depends on an instance of the Pell equation possessing prohibitively large solutions.[1] However, even if the attribution to Archimedes is correct, there is no evidence that he actually solved it or that he ever considered the Pell equation.

We are on firmer ground if we start the history of the Pell equation with Diophantus of Alexandria, who flourished around 250 A.D. according to the best estimate. His book *Arithmetica* is devoted in part to finding positive rational solutions to polynomial equations. Among these are some special cases of the Pell equation. An example where he gave a general solution is easily described in modern symbolism. Suppose $c + d = a^2$ is a square. Then $a, 1$ is a solution to $X^2 - dY^2 = c$. Let x, y be another positive rational solution to this equation. Then $x \neq a$ and $y \neq 1$, so we can write

$$x = t(y - 1) + a \tag{5}$$

for some t. Note that

$$t = \frac{x - a}{y - 1}$$

is rational if x, y are. A little algebra allows us to express x, y rationally in terms of t, thus yielding a parametrisation of the (rational points on the) curve $X^2 - dY^2 = c$: From (5) and the assumption that x, y satisfy (4), it follows that

$$\left(t(y - 1) + a\right)^2 - dy^2 = c.$$

Thus

$$t^2(y - 1)^2 + 2t(y - 1)a + a^2 - dy^2 = c$$

$$t^2(y - 1)^2 + 2at(y - 1) = dy^2 + c - a^2 = dy^2 + c - c - d$$
$$= d(y^2 - 1) = d(y + 1)(y - 1)$$

$$t^2(y - 1) + 2at = d(y + 1) = d(y - 1) + 2d$$
$$(y - 1)(t^2 - d) = 2d - 2at$$
$$y - 1 = 2 \cdot \frac{d - at}{t^2 - d}.$$

---

[1] A nice discussion of the cattle problem can be found on pages 249 - 252 of the chapter "The Pellian" of Albert H. Beiler, *Recreations in the Theory of Numbers; The Queen of Mathematics Entertains*, Dover Publications, New York, 1966.

Thus

$$y = 2 \cdot \frac{d - at}{t^2 - d} + 1 \tag{6}$$

$$x = 2t \cdot \frac{d - at}{t^2 - d} + a. \tag{7}$$

One can, of course, simplify each of (6), (7), but the important point, that x, y are rationally expressed in terms of t has been made. Every rational solution to (4), where $d + c$ is a perfect square, is given by (6), (7) for some rational value of t. Of course, one must still sort out which values of t yield positive x, y in accordance with Diophantus's wishes, and which yield integral solutions in accordance with modern taste. The first of these tasks is unimportant—if a negative value is obtained for x or y, simply ignore the sign. For, $(-x)^2 = x^2$ and $(-y)^2 = y^2$. For example, if $d = 3$ and $c = 1$, then $3 + 1 = 2^2$ and $a = 2$. Every rational solution to

$$X^2 - 3Y^2 = 1 \tag{8}$$

is of the form

$$x = \frac{-2t^2 + 6t - 6}{t^2 - 3}, \quad y = \frac{t^2 - 4t + 3}{t^2 - 3}$$

for some rational value of t. In particular, the choice $t = 0$ yields $x = -6/-3 = 2, y = 3/-3 = -1$ and we simply ignore the sign of y to obtain the solution $2, 1$.

The problem of deciding which values of t yield integral solutions is harder. For example, for $d = 3$ again, the choice $t = 4$ yields the fractional values $x = -14/13, y = 3/13$ (whence the positive solution $14/13, 3/13$), while $t = 5/3$ yields the integral solution $7, 4$. A different approach is needed.

The next chapter in the history of the Pell equation was written by the Indians. In the 7th century A.D., Brahmagupta observed that one can multiply solutions of the Pell equations (4) for fixed d and given values c_1, c_2 of c to obtain a solution for the same d and $c = c_1 c_2$. The method of multiplication is very simple. To any Pellian expression $x^2 - dy^2$, with x, y rational, one associates the quadratic irrational $x + y\sqrt{d}$. Given two solutions to (4) with common d, but possibly different c's,

$$x_1^2 - dy_1^2 = c_1, \quad x_2^2 - dy_2^2 = c_2,$$

one multiplies the associated quadratic irrationals:

$$(x_1 + y_1\sqrt{d})(x_2 + y_2\sqrt{d}) = (x_1 x_2 + dy_1 y_2) + (x_1 y_2 + x_2 y_1)\sqrt{d},$$

and observes

$$(x_1 x_2 + dy_1 y_2)^2 - d(x_1 y_2 + x_2 y_1)^2$$
$$= \left((x_1 x_2 + dy_1 y_2) + (x_1 y_2 + x_2 y_1)\sqrt{d}\right)\left((x_1 x_2 + dy_1 y_2) - (x_1 y_2 + x_2 y_1)\sqrt{d}\right)$$
$$= (x_1 + y_1\sqrt{d})(x_2 + y_2\sqrt{d})(x_1 - y_1\sqrt{d})(x_2 - y_2\sqrt{d})$$

$$= \left(x_1 + y_1\sqrt{d}\right)\left(x_1 - y_1\sqrt{d}\right)\left(x_2 + y_2\sqrt{d}\right)\left(x_2 - y_2\sqrt{d}\right)$$
$$= \left(x_1^2 - dy_1^2\right)\left(x_2^2 - dy_2^2\right) = c_1 c_2.$$

If one starts with a single integral solution to (3),

$$x^2 - dy^2 = 1,$$

multiplying the associated quadratic irrational by itself produces another, larger integral solution,

$$\left(x^2 + dy^2\right)^2 - d(2xy)^2 = 1 \cdot 1 = 1.$$

If one starts with an integral solution to (4) with $c = -1$, multiplying the associated irrational by itself will yield an integral solution to (3):

$$\left(x^2 + dy^2\right)^2 - d(2xy)^2 = (-1) \cdot (-1) = 1.$$

An integral solution to (4) for $c = \pm 2$, upon being multiplied by itself, yields a solution for $c = 4$, from which one for $c = 1$ is quickly obtained. From

$$x^2 - dy^2 = \pm 2,$$

we have

$$(x^2 + dy^2)^2 - d(2xy)^2 = (\pm 2)^2 = 4.$$

Assuming x, y are integers,

$$(x^2 + dy^2)^2 = d(2xy)^2 + 4$$

is even, whence $x^2 + dy^2$ is even and

$$\left(\frac{x^2 + dy^2}{2}\right)^2 - d(xy)^2 = 1,$$

whence $(x^2 + dy^2)/2, xy$ is an integral solution to (3).

1 Exercise. From the following solutions to (4) derive new integral solutions to (3):
i. $(d = 3)\ 7^2 - 3 \cdot 4^2 = 1$
ii. $(d = 5)\ 2^2 - 5 \cdot 1^2 = -1$
iii. $(d = 2)\ 4^2 - 2 \cdot 3^2 = -2.$

For a fixed d, it is easy to find solutions x, y, c to (4). And from such one can try to derive a solution to (3). Take Brahmagupta's example (2):

$$X^2 - 92Y^2 = 1.$$

If we let $y = 1$, $92y^2$ is just 92 and $100 = 10^2$ is the closest square. Thus

$$10^2 - 92 \cdot 1^2 = 8.$$

Multiplying $10 + 1 \cdot \sqrt{92}$ by itself yields

$$x = 10^2 + 92 \cdot 1 = 192, \quad y = 10 \cdot 1 + 1 \cdot 10 = 20$$

is a solution to

$$X^2 - 92Y^2 = 8^2. \qquad (9)$$

Both x and y are divisible by 4, so dividing (9) by 4^2 yields

$$\left(\frac{x}{4}\right)^2 - 92 \left(\frac{y}{4}\right)^2 = 2^2,$$

i.e.,

$$48^2 - 92 \cdot 5^2 = 4.$$

Another self-multiplication yields

$$(48^2 + 92 \cdot 5^2)^2 - 92(2 \cdot 48 \cdot 5)^2 = 4^2,$$

i.e.,

$$4604^2 - 92 \cdot 480^2 = 4^2,$$

with each x, y clearly divisible by 4. Dividing the equation by 4^2 yields

$$\left(\frac{4604}{4}\right)^2 - 92 \left(\frac{480}{4}\right)^2 = 1,$$

i.e.,

$$1151^2 - 92 \cdot 120^2 = 1.$$

The solution 1151, 120 to Brahmagupta's equation is, in fact, the smallest nontrivial solution to his equation.

One is not always so lucky and Brahmagupta's method was superceded in the 11th century by one described by Ācārya Jayadeva, whose method was superceded the following century by one described by Bhāskara II (b. 1115). Although these procedures are completely general, the proof that they were was only first given in 1929. The method is simple enough, but the motivation is lacking. André Weil[2] (1906 - 1998) sketches the method briefly and offers a glimpse of insight into its workings. To solve (3), one can begin with some solution x, y to (4) for some c. In practice, one does this by choosing $y = 1$, and finding x so that x^2 is close to d. Then one chooses c accordingly:

$$c = x^2 - d \cdot 1^2.$$

We may assume, as in the beginning case with $y = 1$, that x, y are relatively prime. If they are not, they have a greatest common divisor k, and we can divide the equation

$$x^2 - dy^2 = c$$

---

[2] André Weil, *Number Theory: An Approach Through History; From Hammurapi to Legendre*, Birkhäuser Boston, Inc., Boston, 1984, pp. 22 - 24.

by k^2 to obtain

$$\left(\frac{x}{k}\right)^2 - d\left(\frac{y}{k}\right)^2 = \frac{c}{k^2},$$

with $x/k, y/k, c/k^2$ integral and $x/k, y/k$ relatively prime.

One wishes to construct first a solution x', y' to the related equation

$$X^2 - dY^2 = cc',$$

for some c' and multiply the solutions together. If $y' = 1$, the product solution is

$$x'' = xx' + dy, \quad y'' = x \cdot 1 + yx',$$

and it satisfies

$$(x'')^2 - d(y'')^2 = c \cdot cc' = c^2 c'. \tag{10}$$

x, y being assumed relatively prime, it is possible to choose an integer x' so that $y'' = x + yx'$ is divisible by c. This means c^2 divides $c^2 c' + d(y'')^2 = (x'')^2$ and we have

$$\left(\frac{xx' + dy}{c}\right)^2 - d\left(\frac{x + yx'}{c}\right)^2 = c',$$

with

$$\frac{xx' + dy}{c}, \quad \frac{x + yd}{c}, \quad c'$$

integral.

[The existence of c' (i.e., that c divides $(x')^2 - d(y')^2$) was assumed. This existence follows, however, from the assumption that c divides $x + yx'$. Writing

$$x + yx' = cz,$$

we have

$$(yx')^2 = (cz - x)^2 = c^2 z^2 - 2czx + x^2$$

and

$$\begin{aligned}
(x')^2 - d(y')^2 &= (x')^2 - d \\
&= \frac{c^2 z^2 - 2czx + x^2}{y^2} - \frac{x^2 - c}{y^2} \\
&= \frac{c^2 z^2 - 2czx + x^2 - x^2 + c}{y^2} \\
&= c \cdot \frac{cz^2 - 2zx + 1}{y^2}.
\end{aligned}$$

Because x, y are relatively prime and $x^2 - dy^2 = c$, y and c are relatively prime and y^2 divides $cz^2 - 2zx + 1$, making $(x')^2 - d = c \cdot$ integer.]

Having explained all of this, here is the procedure. Start with a nice solution x, y to (4) for some c. Assume y'' is of the form $x + ry$ for some r. Choose r so that $x + ry$ is divisible by c. For reasons not yet explained, choose r so that the difference $r^2 - d$ is as small as possible. c will divide $r^2 - d$ and we can choose $c' = (r^2 - d)/c$. Then there will be a number x'' such that

$$(x'')^2 - d(y'')^2 = c'.$$

I illustrate this with Brahmagupta's example of $d = 92$: As before, we start with $x = 10, y = 1, c = 8$:

$$10^2 - 92 \cdot 1^2 = 8.$$

We want r such that 8 divides $10 + r \cdot 1$ and $r^2 - 92$ is as small as possible. The candidates for r are $-2, 6, 14, \ldots$ Among these $r^2 - 92$ is minimised for $r = 6$. Our new values are

$$y'' = \frac{x + ry}{c} = \frac{10 + 6 \cdot 1}{8} = \frac{16}{8} = 2$$

$$c' = \frac{r^2 - 92}{c} = \frac{36 - 92}{8} = -7$$

$$x'' = \sqrt{c' + d(y'')^2} = \sqrt{-7 + 92 \cdot 2^2} = 19.$$

We now start over with $x = 19, y = 2, c = -7$:

$$19^2 - 92 \cdot 2^2 = -7.$$

We want r such that -7 divides $19 + 2r$ and $r^2 - 92$ is small. The candidates for r are $1, 8, 15, \ldots$ The minimisation occurs at $r = 8$.

$$y'' = \frac{19 + 8 \cdot 2}{-7} = -5$$

$$c' = \frac{8^2 - 92}{-7} = 4$$

$$x'' = \sqrt{4 + 92 \cdot 5^2} = \sqrt{2304} = 48.$$

This gives us

$$48^2 - 92 \cdot 5^2 = 4.$$

We can now take the easy way out like we did before in discussing Brahmagupta's solution to the problem, or continue. Continuing would yield successively

$$211^2 - 92 \cdot 22^2 = -7$$
$$470^2 - 92 \cdot 49^2 = 8$$

and finally

$$1151^2 - 92 \cdot 120^2 = 1.$$

That this procedure finally yields a solution is not at all clear. I shall not attempt to show that it does here. I note merely that it is to guarantee this that r was always chosen to minimise $r^2 - d$.

After going through all of this one might be left thinking it would be easier to simply search for a solution by calculating successively

$$92 \cdot 1^2 + 1, 92 \cdot 2^2 + 1, 92 \cdot 3^2 + 1, \ldots$$

until one of these values turns out to be a perfect square. Obviously, this is not easier if one does one's calculations with pencil and paper, but with today's programmable calculators it must surely be less time consuming to take the trouble of writing a program. Here is a simple one I wrote for the *TI-83 Plus*:

```
PROGRAM:PELL
:Disp "ENTER D"
:Input D
:Disp "ENTER N"
:Input N
:For(I,1,N)
:D*I² + 1 →K
:√(K)→E
:If E = int(E)
:Disp I
:End
```

When executed, the program tells the user to enter a value d to determine which Pell equation (3) is to be solved. It then demands an upper bound n on the number of steps one is to take in a search. Once it has this information it goes through the list $d \cdot 1^2 + 1, d \cdot 2^2 + 1, \ldots$, displaying y every time $dy^2 + 1$ is a perfect square. If there are several solutions below n, they will all be displayed on separate lines followed by "Done", which is displayed when the program finishes execution. The sequence of y's for which $dy^2 + 1$ is a perfect square grows exponentially, so there is little danger of the list of displayed numbers scrolling off the screen.

One can modify the program by having it halt execution after the first solution is obtained. Or one could have it display both x and y (the values of the variables E and I) instead of just y. Also, one could replace the For command by a While or Repeat, thus eliminating the need for an estimated upper bound n on the size of y. One should, however, check on how to stop program execution on one's calculator should the program seem to run forever. (On the *TI-83 Plus*, one simply hits the On/Off button.) In theory, because y exists, the program will eventually find it and the program execution will end. The reality, however, is that the calculator can only handle numbers with so many digits. When this is exceeded, any number of things can go wrong. However, for $d = 92$, the program runs beautifully. If one enters any number $n \geq 120$ for N, the program will display 120 after a few seconds. If the number n is much larger than 120, the program will continue running for a while as it looks for other solutions.

One might imagine that, for small values of d, the program offers the quickest approach. This is not the case. It doesn't work for $d = 61$. The solution, which can be found using paper and pencil in a reasonable amount of time by

Bhāskara's method, involves multiplications resulting in numbers that have too many digits for the calculator. One would need a more sophisticated program dealing with larger numbers represented, say, as lists of smaller ones.

Before putting the calculator away, however, one might like to use it to explore the use of the Pell equation to estimate square roots. Starting with

$$3^2 - 2 \cdot 2^2 = 9 - 8 = 1,$$

we have

$$\frac{3}{2} = 1.5$$

as an approximation to $\sqrt{2}$. Now

$$(3 + 2\sqrt{2})^2 = 17 + 12\sqrt{2},$$

whence

$$\frac{17}{12} = 1.41\overline{6},$$

is a better approximation.

$$(3 + 2\sqrt{2})^3 = (17 + 12\sqrt{2})(3 + 2\sqrt{2}) = 99 + 70\sqrt{2}$$

yields

$$\frac{99}{70} \approx 1.414285714,$$

which is even better. One more step yields

$$(3 + 2\sqrt{2})^4 = (99 + 70\sqrt{2})(3 + 2\sqrt{2}) = 577 + 408\sqrt{2},$$

and

$$\frac{577}{408} \approx 1.414215686$$

is a further improvement.

2 Exercise. Find some simple approximations to $\sqrt{3}$ or $\sqrt{5}$ by considering the solutions to

$$X^2 - 3Y^2 = 1 \quad \text{or} \quad X^2 - 5Y^2 = 1,$$

respectively.

The Pell equation enters modern mathematics with Pierre de Fermat in the 17th century. Fermat had read Diophantus and become interested in number theory and tried without much success to interest his fellow mathematicians in the subject. In February 1657, he proposed the problem of solving the Pell equation in integers as a challenge to the English mathematicians via Sir Kenelm Digby (1603 - 1665), a nobleman and minor figure in the history of mathematics. The copy of the letter that was prepared for Lord Brouncker (1620 - 1684) by Digby's secretary, however, omitted the passage demanding that the solutions be integral. Brouncker thoroughly solved the problem in rationals. Fermat was unimpressed, for the solution in rationals is fairly simple.

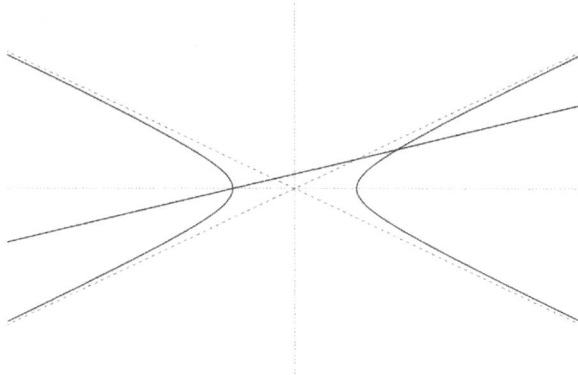

FIGURE 1.

A slight variation on Brouncker's approach can be illustrated graphically. Consider the hyperbola graphing the equation $X^2 - dY^2 = 1$. Any line passing through $(-1, 0)$ with slope t between $-1/\sqrt{d}$ and $1/\sqrt{d}$ intersects the right half of the hyperbola in a unique point. (See FIGURE 1.) The nonnegative rational points (x, y) on the hyperbola can thus be parametrically defined as functions of the slope t of the line passing through $(-1, 0)$ for $0 \leq t < 1/\sqrt{d}$. The calculation is fairly straightforward: The equation of the line passing through (x, y) and $(-1, 0)$ is

$$Y - 0 = t(X - (-1))$$

whence

$$y = t(x + 1). \tag{11}$$

But, if (x, y) also lies on the hyperbola, we have

$$x^2 - dy^2 = 1$$
$$x^2 - dt^2(x + 1)^2 = 1$$
$$x^2(1 - dt^2) - 2dt^2 x - dt^2 - 1 = 0,$$

so

$$x = \frac{2dt^2 \pm \sqrt{4d^2t^4 - 4(1 - dt^2)(-1 - dt^2)}}{2(1 - dt^2)}$$

$$= \frac{2dt^2 \pm 2\sqrt{d^2t^4 + (1 - d^2t^4)}}{2(1 - dt^2)}$$

$$= \frac{dt^2 \pm 1}{1 - dt^2}.$$

Thus x is either

$$\frac{1 + dt^2}{1 - dt^2} \quad \text{or} \quad \frac{dt^2 - 1}{1 - dt^2} = -1.$$

The second is on the wrong branch of the hyperbola and we have

$$x = \frac{1 + dt^2}{1 - dt^2}. \tag{12}$$

Together with (11), this yields

$$y = t\left(\frac{1 + dt^2}{1 - dt^2} + 1\right) = t\left(\frac{1 + dt^2 + 1 - dt^2}{1 - dt^2}\right)$$
$$= \frac{2t}{1 - dt^2}. \tag{13}$$

For the purpose of approximating $\sqrt{d}$, Brouncker's solution is fine. One has

$$\sqrt{d} \approx \frac{x}{y} = \frac{1 + dt^2}{2t}.$$

We assume here that

$$0 < t < 1/\sqrt{d}.$$

A little algebra shows that, for such t,

$$\sqrt{d} < \frac{x}{y} < \frac{1}{t}$$

and x/y is a better approximation to $\sqrt{d}$ than $1/t$. Letting $u = 1/t$, we can describe the process simply. Start with any rational number $u > \sqrt{d}$ and improve the estimate by choosing[3]

$$u' = \frac{1 + d\left(\frac{1}{u}\right)^2}{2\left(\frac{1}{u}\right)} = \frac{u^2 + d}{2u}.$$

For $d = 2$, for example, if we start with $u = 2$, we get

$$u' = \frac{4 + 2}{4} = \frac{3}{2}, \quad u'' = \frac{\frac{9}{4} + 2}{2 \cdot \frac{3}{2}} = \frac{9 + 8}{4 \cdot 3} = \frac{17}{12}$$

$$u''' = \frac{\left(\frac{17}{12}\right)^2 + 2}{2 \cdot \frac{17}{12}} = \frac{17^2 + 2 \cdot 12^2}{2 \cdot 17 \cdot 12} = \frac{577}{408},$$

which, curiously enough, is the same sequence we had before. For $d = 92$, we could start with $u = 10$ and get

$$u' = \frac{100 + 92}{2 \cdot 10} = \frac{192}{20} = \frac{48}{5}$$

---

[3] This method of improving one's approximation to the square root goes back to the Greeks—without reference to the Pell equation. Cf. Chapter 8, below.

$$u'' = \frac{\left(\dfrac{48}{5}\right)^2 + 92}{2 \cdot \dfrac{48}{5}} = \frac{48^2 + 92 \cdot 25}{2 \cdot 48 \cdot 5} = \frac{4604}{480} = \frac{1151}{120}.$$

Again, $1151, 120$ is, oddly enough, the smallest integral solution to (2).

For Fermat's purpose of finding the integral solutions to the Pell equation, Brouncker's method is useless. Now, it is not hard to see that x, y are rational exactly when t is rational. So one might try setting $t = m/n$, with m, n whole numbers and see if one can determine when x, y are whole numbers:

$$x = \frac{1 + d\left(\dfrac{m}{n}\right)^2}{1 - d\left(\dfrac{m}{n}\right)^2} = \frac{n^2 + dm^2}{n^2 - dm^2}$$

$$y = \frac{2\left(\dfrac{m}{n}\right)}{1 - d\left(\dfrac{m}{n}\right)^2} = \frac{2mn}{n^2 - dm^2}.$$

Now x, y are integral just in case $n^2 - dm^2$ divides both $n^2 + dm^2$ and $2mn$, in particular if it divides the greatest common divisor of the two numbers. Assuming that m/n in lowest terms, and that d is square-free (i.e., d has no repeated factors), this common divisor can be shown to divide 2. Thus we must have

$$n^2 - dm^2 = 1 \quad \text{or} \quad n^2 - dm^2 = 2$$

and we can find x, y integral solutions to (3) if we can find integral solutions to the Pell equation (3) or its similar companion

$$X^2 - dY^2 = 2. \tag{14}$$

For many values of d (e.g., $d = 5$) equation (14) has no solution and we see that, for such d, Brouncker's solution reduces the problem of solving (3) in integers to the problem of solving (3) in integers, i.e., the problem is reduced to itself and nothing has been gained.

Brouncker returned to the problem and almost completely solved it. To state succinctly what he did, we define the *fundamental solution* x, y to (3) to be a solution in which x, y are positive integers and for which no solution in positive integers u, v exists for u or v smaller than x or y, respectively. Brouncker showed how to find the fundamental solution and how to generate all integral solutions from it. He did not prove, however, that an integral solution always exists, but assumed this.

A great deal of activity on the Pell equation ensued. The high points were achieved by Euler (1707 - 1783), Lagrange, and finally Dirichlet (1805 - 1859). Euler came a few mathematical generations after Fermat and Brouncker. These latter two men lived on the eve of the Calculus, followed immediately by Newton and Leibniz and their discovery of the algorithms that gave the Calculus its

The Bernoulli family is one of the great scientific dynasties and any work on the history of mathematics in the 17th and 18th centuries is bound to mention more than one member of the family. As some names are repeated and their names often appear translated, some confusion can arise. Thus I shall attempt to sort them out. A druggist named Jakob Bernoullis moved from Amsterdam to Basel. His son Nikolaus was an artist who had three sons: Jakob (1654 - 1705), Nikolaus, and Johann (1667 - 1748). Jakob and Johann are the famous feuding brothers, and are cited in the literature under various names. Jakob is also referred to as Jacques, James, and, as he was the first Bernoulli bearing the name, Jakob I. Johann is likewise variously referred to as Jean, John, and Johann I. Nikolaus was a painter, but he had a son likewise named Nikolaus (1687 - 1759) who was a doctor of law as well as a mathematician. He is often referred to as Nikolaus I Bernoulli. Jakob I had a son Nikolaus who, like his namesake uncle and grandfather, was an artist. Johann I was more fortunate in his offspring, siring three mathematicians: Nikolaus II (1695 - 1726), Daniel I (1700 - 1782), and Johann II (1710 - 1770). Johann II had three scientific sons, Johann III (1744 - 1807) who was a mathematician and astronomer, Daniel II (1757 - 1834) who was a medical doctor and assisted Daniel I, and Jakob II (1759 - 1789) who was another mathematician. This ended the mathematical chapter of the Bernoulli family history. Later Bernoullis distinguished themselves in other pursuits. In the next generation, for example, one finds Christoph Bernoulli (1782 - 1863), the son of Daniel II, being a professor of Natural History. The three most famous Bernoullis are Jakob I, Johann I, and Daniel I.

name. Two of Leibniz's associates were the brothers Bernoulli, Jakob (1654 - 1705) and Johann (1667 - 1748). Euler was a student of Johann.

Leonhard Euler was one of the most prolific mathematicians of all time, his work touching on all aspects of the subject. He made several contributions to the study of the Pell equation. One, which we will consider later, was to observe that if d is of the form $a^2 - 1$, the fundamental solution of its Pell equation is given by $x = a, y = 1$, and if d is of the form $a^2 + 1$, its fundamental solution is given by $x = 2a^2 + 1, y = 2a$. That is, he observed the identities[4],

$$a^2 - (a^2 - 1) \cdot 1^2 = 1 \tag{15}$$

---

[4] Leonhard Euler, *Elements of Algebra*, Springer-Verlag, New York, nd, pp. 352 - 359. This is a reprint of the English translation by Rev. John Hewlett of Bernoulli's French translation, with additions by Joseph Louis Lagrange on continued fractions and additional algebraic topics. The title page announces the 1840 publication to be the 5th edition, though it is not clear if this means it is the 5th edition of the English translation, the translation of the 5th edition of the French translation, or the translation of the French translation of the 5th German edition. It is also nowhere made clear which of the Bernoullis was the French translator. (I've read that it was Johann III.) Originally published in 1770, Euler's algebra book is second only to Euclid's *The Elements* as a textbook of mathematics in longevity and influence.

$$(2a^2 + 1)^2 - (a^2 + 1)(2a)^2 = 1 \tag{16}$$

$$(a^2 - 1)^2 - (a^2 - 2)a^2 = 1 \tag{17}$$

$$(a^2 + 1)^2 - (a^2 + 2)a^2 = 1. \tag{18}$$

He also noted a few other such solutions, but they are not as simple and I will not repeat them here.

Euler's second major contribution was to rediscover Brahmagupta's multiplication rule. He first published this in 1759, though he mentioned it in a letter to Christian Goldbach (1690 - 1764) already in 1753. Europeans at this stage were unaware of the work of the Indian mathematicians. Such knowledge came in the 19th century through the efforts of Henry Thomas Colebrooke (1765 - 1837). Colebrooke had spent 32 years in India working as a magistrate for the East India Company. He learned Sanskrit so as to be able to read Sanskrit legal books. He published a Sanskrit grammar in 1805 and a dictionary in 1808. His most important mathematical contribution was his *Algebra, with Arithmetic and Mensuration, from the Sanskrit of Brahmegupta and Bhāscara*, published in London in 1817.[5]

The late translation of Brahmagupta and Bhāskara into a European language explains the duplication of effort by the Europeans and their failure to credit the Indians. A further misassignment of credit was made by Euler[6], who misunderstood some passage in a work of John Wallis (1616 - 1703) and credited Brouncker's solution to John Pell (1611 - 1685). The name has stuck and expanded to cover a number of "Pellian" constructs. Eponymous names in mathematics should never be taken too seriously, but should be treated as at best rough first approximations to a proper attribution of credit.

Getting back to Euler: His third main contribution to the Pell equation was to express Brouncker's algorithm for generating the fundamental solution to (3) in terms of the continued fraction expansion of $\sqrt{d}$.

A *continued fraction* is an expression of the form

$$a_0 + \cfrac{b_0}{a_1 + \cfrac{b_1}{a_2 + \cfrac{b_2}{a_3 + \ldots}}}$$

The meaning attached to such an infinite expression is the limit of the sequence of *convergents*,

$$a_0, \quad a_0 + \frac{b_0}{a_1}, \quad a_0 + \cfrac{b_0}{a_1 + \cfrac{b_1}{a_2}}, \ldots$$

---

[5] Joseph Dauben and Christoph Scriba eds., *Writing the History of Mathematics: Its Historical Development*, Birkhäuser Verlag, Basel, 2002, pp. 398 - 399. This is a rather old-fashioned book, written in the style of many old histories of mathematics, describing the contributions of individual historians, country by country—each history written by a different historian or group of historians. The second half of the volume is a biographical dictionary of historians of mathematics.

[6] *Op. cit.*, p. 352.

In general the a_i's and b_i's may be arbitrary real numbers. In the most common case, all the b_i's are 1, all the a_i's are integers, positive for $i > 0$. Continued fractions of this form are called *simple continued fractions*.

Already in 1572 Rafael Bombelli (1526 - 1572) had essentially found the continued fraction expansion

$$\sqrt{13} = 3 + \cfrac{4}{6 + \cfrac{4}{6 + \ldots}}$$

And in 1613 Pietro Antonio Cataldi (1552 - 1626) had noted

$$\sqrt{18} = 4 + \cfrac{2}{8 + \cfrac{2}{8 + \ldots}}$$

Continued fractions entered the mainstream of mathematics with Lord Brouncker, who showed around 1858 that

$$\frac{4}{\pi} = 1 + \cfrac{1}{2 + \cfrac{9}{2 + \cfrac{25}{2 + \cfrac{49}{2 + \cfrac{81}{2 + \ldots}}}}}$$

[With regard to π, Lambert (1728 - 1777) determined in 1770 that

$$\pi = 3 + \cfrac{1}{7 + \cfrac{1}{15 + \cfrac{1}{1 + \cfrac{1}{292 + \ldots}}}}$$

The first few convergents are

$$3, \qquad 3 + \frac{1}{7} = \frac{22}{7}, \qquad 3 + \cfrac{1}{7 + \cfrac{1}{15}} = 3 + \frac{15}{105 + 1} = \frac{318 + 15}{106} = \frac{333}{106}$$

and

$$3 + \cfrac{1}{7 + \cfrac{1}{15 + \cfrac{1}{1}}} = 3 + \cfrac{1}{7 + \cfrac{1}{16}} = 3 + \frac{16}{112 + 1} = \frac{355}{113}.$$

The second estimate on the list, $22/7$, is the familiar Archimedean value. The last, due originally to Zŭ Chōngzhī, based on the sequence 113355, is particularly easy to remember and is quite accurate.]

As already mentioned, Brouncker's method of finding the fundamental solution to the Pell equation is essentially a continued fraction expansion. Lagrange would later prove that the simple continued fraction expansion of $\sqrt{d}$ is ultimately periodic and the fundamental solution is given by the numerator and denominator of the convergents occurring at the end of the first or second period. The basic theory does not go much beyond high school mathematics: some basic algebra and, at some point, the notion of limit, are all that is needed. A good elementary exposition is C.D. Olds, *Continued Fractions*, Random House, 1963.

We are not going to worry here about theory and justification. I wish only to offer a primitive illustration of their use.

From our familiar $10^2 - 92 \cdot 1^2$, we get $92 = 10^2 - 8$, which we might think of as $10^2 - 4 \cdot 2 = b^2 - 4c$, the discriminant of the quadratic equation,

$$X^2 - 10X + 2 = 0.$$

The larger solution to this is

$$x = \frac{10 + \sqrt{100 - 4 \cdot 2}}{2} = \frac{10 + \sqrt{92}}{2},$$

and we have

$$\sqrt{92} = 2x - 10.$$

Now consider

$$x^2 = 10x - 2$$
$$x = 10 - \frac{2}{x}$$
$$= 10 - \cfrac{2}{10 - \cfrac{2}{x}}$$
$$= 10 - \cfrac{2}{10 - \cfrac{2}{10 - \cfrac{2}{10 - x}}}$$
$$= \text{etc.}$$

This expresses

$$\frac{10 + \sqrt{92}}{2} = x = 10 - \cfrac{2}{10 - \cfrac{2}{10 - \cfrac{2}{10 - \dots}}}$$

as an infinite continued fraction. The first convergent is

$$x \approx 10$$

with associated estimate

$$2x - 10 \approx 2 \cdot 10 - 10 = 10 = \frac{10}{1}$$

for $\sqrt{92}$. And we note that

$$10^2 - 92 \cdot 1^2 = 8.$$

The next round yields

$$x \approx 10 - \frac{2}{10} = \frac{100 - 2}{10} = \frac{98}{2} = \frac{49}{5}$$

$$\sqrt{92} = 2x - 10 \approx 2 \cdot \frac{49}{5} - 10 = \frac{98 - 50}{5} = \frac{48}{5}$$

$$48^2 - 92 \cdot 5^2 = 4.$$

The next round yields

$$x \approx 10 - \frac{2}{49/5} = \frac{490 - 10}{49} = \frac{480}{49}$$

$$\sqrt{92} = 2x - 10 \approx 2 \cdot \frac{480}{49} - 10 = \frac{960 - 490}{49} = \frac{470}{49}$$

$$470^2 - 92 \cdot 49^2 = 8.$$

And again:

$$x \approx 10 - \frac{2}{480/49} = \frac{4800 - 98}{480} = \frac{4702}{480} = \frac{2351}{240}$$

$$\sqrt{92} = 2x - 10 \approx 2 \cdot \frac{2351}{240} - 10 = \frac{2351 - 1200}{120} = \frac{1151}{120}$$

$$1151^2 - 92 \cdot 120^2 = 1.$$

Another example is $d = 13$. If we examine successive values of $13y^2$ and squares close to them,

$$4^2 - 13 \cdot 1^2 = 3$$
$$7^2 - 13 \cdot 2^2 = -3$$
$$11^2 - 13 \cdot 3^2 = 4,$$

we see that $13 \cdot 9 = 11^2 - 4$ is the discriminant of

$$X^2 - 11X + 1 = 0.$$

If x is a solution to this, we have

$$x = \frac{11 + 3\sqrt{13}}{2},$$

whence

$$\sqrt{13} = \frac{2x - 11}{3}.$$

Again we have

$$x = 11 - \frac{1}{x} = 11 - \cfrac{1}{11 - \cfrac{1}{11 - \ldots}}$$

And we can consider the convergents. First, we have

$$x \approx 11$$

$$\sqrt{13} = \frac{2x - 11}{3} \approx \frac{22 - 11}{3} = \frac{11}{3}$$

$$11^2 - 13 \cdot 3^2 = 4.$$

Then

$$x \approx 11 - \frac{1}{11} = \frac{121 - 1}{11} = \frac{120}{11}$$

$$\sqrt{13} \approx \frac{2 \cdot \frac{120}{11} - 11}{3} = \frac{240 - 121}{3 \cdot 11} = \frac{119}{33}$$

$$119^2 - 13 \cdot 33^2 = 4.$$

And

$$x \approx 11 - \frac{1}{120/11} = \frac{1320 - 11}{120} = \frac{1309}{120}$$

$$\sqrt{13} \approx \frac{2 \cdot \frac{1309}{120} - 11}{3} = \frac{\frac{1309}{60} - \frac{660}{60}}{3} = \frac{649}{180}$$

$$649^2 - 13 \cdot 180^2 = 1.$$

The solution isn't always this easy, and in these examples I've used, not the continued fraction expansions of $\sqrt{d}$, but the expansions of some auxiliary solutions to related equations. In general, however, one can use a continued fraction expansion of $\sqrt{d}$ and at some point the numerator and denominator of a convergent will yield the fundamental solution. Euler was explicit about this and showed that this was, in effect, what Brouncker had been doing. Eventually Lagrange proved that the simple continued fraction expansions of quadratic irrationals were ultimately periodic and that the method did yield the fundamental solution to the Pell equation.

It was not until 1929 that it was proven that the method given by Bhāskara always worked and, although it was oft repeated that it too was a continued fraction expansion, this was not proven until several more decades had passed. Bhāskara's method reduces to a continued fraction expansion, but one with more general (e.g., negative) entries than those of Lagrange's simple continued fractions. His method is generally more efficient than the European one, with smaller periods and smaller numbers involved.

Lagrange did his work in the late 1700's. Starting in the 1830's, Peter Gustav Lejeune Dirichlet studied the Pell equation from a different angle. His proof, as presented by Richard Dedekind (1831 - 1916) in editing Dirichlet's treatise on number theory, has a lemma and three steps. The lemma is Dirichlet's famous *Pigeon Hole Principle.*

3 Theorem (Pigeon Hole Principle). *If one distributes n objects into m containers and $n > m$, some container will receive more than one object.*

The Pigeon Hole Principle is applied at various points in the proof. First it is used to show that, for any irrational number ξ, there are infinitely many positive integers x, y such that

$$0 < \frac{x}{y} - \xi < \frac{1}{y^2}.$$

The next step is to observe that, for $\xi = \sqrt{d}$ and such x, y, one has

$$0 < x^2 - dy^2 < 1 + 2\sqrt{d}.$$

There are only finitely many values c with $0 < c < 1 + 2\sqrt{d}$ and infinitely many pairs x, y satisfying

$$0 < x^2 - dy^2 < 1 + 2\sqrt{d},$$

so for some value of c there are infinitely many x, y such that

$$x^2 - dy^2 = c. \tag{19}$$

The idea now is to take two solutions x_1, y_1 and x_2, y_2 to (19) and divide the larger by the smaller:

$$
\begin{aligned}
u + v\sqrt{d} &= (x_1 + y_1\sqrt{d})/(x_2 + y_2\sqrt{d}) \\
&= (x_1 + y_1\sqrt{d})(x_2 - y_2\sqrt{d}) \cdot \frac{1}{c} \\
&= \frac{x_1 x_2 - dy_1 y_2}{c} + \frac{y_1 x_2 - x_2 y_1}{c}\sqrt{d}.
\end{aligned}
$$

The numbers u, v satisfy

$$u^2 - dv^2 = (x_1^2 - dy_1^2)(x_2^2 - dy_2^2) \cdot \frac{1}{c^2} = c \cdot c \cdot \frac{1}{c^2} = 1,$$

by Brahmagupta's multiplication rule. All that remains is to verify that one can choose x_1, y_1 and x_2, y_2 so that c divides $x_1 x_2 - dy_1 y_2$ and $y_1 x_2 - x_1 y_2$. To this end, for any such pair, x, y let

$$a = \text{ the remainder of } x \text{ on division by } c$$
$$b = \text{ the remainder of } y \text{ on division by } c.$$

There are only c^2 pairs a, b and infinitely many pairs x, y with $x^2 - dy^2 = c$. Hence at least two of them, say x_1, y_1 and x_2, y_2 have the same remainders a, b. So we have

$$x_2 = x_1 + ic, \quad y_2 = y_1 + jc,$$

for some integers i, j. But

$$\begin{aligned}
x_1 x_2 - d y_1 y_2 &= x_1(x_1 + ic) - d y_1(y_1 + jc) \\
&= x_1^2 + x_1 ic - d y_1^2 - d y_1 jc \\
&= x_1^2 - d y_1^2 + c(x_1 i - d y_1 j) \\
&= c + c(x_1 i - d y_1 j)
\end{aligned}$$

is a multiple of c. Similarly,

$$\begin{aligned}
y_1 x_2 - x_1 y_2 &= y_1(x_1 + ic) - x_1(y_1 + jc) \\
&= x_1 y_1 + i c y_1 - x_1 y_1 - x_1 jc \\
&= c(i y_1 - j x_1),
\end{aligned}$$

another multiple of c.

The details I've omitted are all as elementary and easily followed. The overall strategy of the proof is inspired and takes a bit of study to understand, but is is the simplest existence proof for integral solutions to the Pell equation there is. Computationally, however, it is not much use. Studying the proof carefully, one can extract an upper bound B on the sizes of x and y, but it is a bit too large to make a simple search for the solution practical. Dirichlet's proof is, however, useful in theoretical matters—it generalises in a couple of different directions to yield fundamental results in higher number theory.

This is about as far as we can go here into the mysteries and history of the Pell equation. I'd like to finish up with a few comments on Euler's identities (15) - (18) and some of their variants. First let me cite a few references. When I was a student there was a publishing company named Chelsea that published inexpensive, high quality mathematics books. These were usually classics that were out of copyright or German books the copyrights of which had been seized by the US Attorney General and assigned to American publishers following Germany's defeat in the Second World War. One book Chelsea announced, but sadly never published, was an omnibus edition of two books on the history of the Pell equation:

H. Konen, *Geschichte der Gleichung* $t^2 - Du^2 = 1$, Leipzig, 1901.

Edward Everett Whitford, *The Pell Equation*, New York, 1912.

Today, I am happy to report, both books are available online and Whitford's book has been published in an inexpensive paperback edition by Merchant Books (2008). It is full of information, even quoting Fermat's letter to Digby in full[7], and finishes with a 40 page mildly annotated bibliography of works on the Pell equation that appeared between 1798 and 1911.

---

[7] Whitford's translation of Fermat's letter also appears in David Eugene Smith, *A Source Book in Mathematics*, 1929, currently published by Dover Publications, Inc. It appears on pp. 215 - 216 of the first volume of the 1959 two-volume reprint.

For the history of the Indian contributions to the study of the Pell equation, I recommend specialised scholarly papers. A bit old, but valuable in this respect is

A.A. Krishnaswamy Ayyangar, "New light on Bhaskara's chakravala or cyclic method of solving indeterminate equations of the second degree in two variables", *Journal of the Indian Math. Society*, series 1, volume 18, part 2 (1929/1930), pp. 225 - 248.

The mathematical portions of the paper are not for the mathematically squeamish, but the general discussion is good. This paper, incidentally, is the work in which was given the first proof that the Indian algorithm always works.

I might also mention a more mathematical, less historical work,

Edward J. Barbeau, *Pell's Equation*, Springer Verlag, New York, 2003.

I've not examined the book carefully, but my impression from glancing at a few pages online is that it is a fair exposition of the Pell equation and its applications accessible to the mathematics undergraduate.

Let us return now to Euler's identities (15) and (16). Identities like these are solutions to Pell equations in the set of polynomials: Given a nonsquare polynomial with integral coefficients $D(X)$, can one find polynomials $P(X), Q(X)$ with integral coefficients such that

$$P(X)^2 - D(X)Q(X)^2 = 1? \tag{20}$$

The last time I looked into it, almost 20 years ago, the problem had not been studied much. But a few things were known. First, that $D(X)$ not be a square is not a sufficient condition. Most important, when there is a solution, there is a fundamental one, i.e., there are P^*, Q^* of minimum degree, with P^* nonconstant, satisfying (20) and such that all other solutions P, Q are of the form

$$P(X) + Q(X)\sqrt{D(X)} = (P^*(X) + Q^*(X)\sqrt{D(X)})^n$$

for some n. Moreover, there are infinitely many polynomials $D(X)$ for which such $P(X), Q(X)$ can be found. Of particular interest for our little study here is the fact that, for every nonsquare value of d, a quadratic polynomial $D(X) = aX^2 + bX + c$ can be found for which there is an integer m such that $D(m) = d$, and, if $P(X), Q(X)$ is the fundamental solution to (20), then $P(m), Q(m)$ is the fundamental solution to $X^2 - dY^2 = 1$.

If $D(X) = aX^2 + bX + c$ is quadratic for which P, Q satisfying (20) exist, then a must be a perfect square, and via a substitution $X = Y + k$, it can be brought into the form $D(Y) = aY^2 + b'Y + c'$ with $0 \le b' \le a$. We thus consider $D(X) = aX^2 + bX + c$ to be in *reduced* form if $0 \le b \le a$. For any such a, b, there are only finitely many values of c—if any— for which solutions $P(X), Q(X)$ to (20) exist, and they can be tabulated. For $a = 1$, we collect the results together in TABLE 1.

The next possible value of a is 4. There are 16 such D's which I collect together in TABLE 2. They are listed slightly differently because of the size of a couple of the polynomials P, Q.

$D(X)$	$P(X)$	$Q(X)$
$X^2 - 1$	X	1
$X^2 + 1$	$2X^2 + 1$	$2X$
$X^2 - 2$	$X^2 - 1$	X
$X^2 + 2$	$X^2 + 1$	X
$X^2 + X$	$2X + 1$	2

TABLE 1. Simple Polynomial Solutions

$D(X) = 4X^2 - 1$
$P(X) = 8X$
$Q(X) = 4$

$D(X) = 4X^2 + 1$
$P(X) = 8X^2 + 1$
$Q(X) = 4X$

$D(X) = 4X^2 - 2$
$P(X) = 4X^2 - 1$
$Q(X) = 2X$

$D(X) = 4X^2 + 2$
$P(X) = 4X^2 + 1$
$Q(X) = 2X$

$D(X) = 4X^2 - 4$
$P(X) = 2X^2 - 1$
$Q(X) = X$

$D(X) = 4X^2 + 4$
$P(X) = 2X^2 + 1$
$Q(X) = X$

$D(X) = 4X^2 - 8$
$P(X) = 2X^4 - 4X^2 + 1$
$Q(X) = X^3 - X$

$D(X) = 4X^2 + 8$
$P(X) = 2X^4 + 4X^2 + 1$
$Q(X) = X^3 + X$

$D(X) = 4X^2 + X$
$P(X) = 8X + 1$
$Q(X) = 4$

$D(X) = 4X^2 + 2X$
$P(X) = 4X + 1$
$Q(X) = 2$

$D(X) = 4X^2 + 4X$
$P(X) = 2X + 1$

$$Q(X) = 1$$
$$D(X) = 4X^2 + 4X - 1$$
$$P(X) = 4X^2 + 4X$$
$$Q(X) = 2X + 1$$

$$D(X) = 4X^2 + 4X + 2$$
$$P(X) = 8X^2 + 8X + 3$$
$$Q(X) = 4X + 2$$

$$D(X) = 4X^2 + 4X - 3$$
$$P(X) = 4X^3 + 6X + 1$$
$$Q(X) = 2X^2 + 2X$$

$$D(X) = 4X^2 + 4X + 3$$
$$P(X) = 4X^2 + 4X + 2$$
$$Q(X) = 2X + 1$$

$$D(X) = 4X^2 + 4X + 5$$
$$P(X) = 32X^6 + 96X^5 + 168X^4 + 176X^3 + 120X^2 + 48X + 9$$
$$Q(X) = 16X^5 + 40X^4 + 56X^3 + 44X^2 + 20X + 4$$

Table 2. More Polynomial Solutions

Somewhere I have a list of all permissible D's for $a = 9, 16, 25, 36, 49, 64, 100$ and the associated P's and Q's for $a = 9, 16$ and 25. These lists are too long to be reproduced here.

If $D(m) = d$, it need not always be the case that $P(m), Q(m)$ yield the fundamental solution to (3), but they usually seem to do so. One can make a table of values of $D(m), P(m), Q(m)$ for the various choices of D and thus get a good start on a table of solutions to (3) for small values of d. The last D listed, $D(X) = 4X^2 + 4X + 5$, is the only one on the list so far given representing $d = 13 \ (= D(1))$, $29 \ (= D(2))$, $53 \ (= D(3))$, or $85 \ (= D(4))$. Brahmagupta's $d = 92$ is comparatively simple: $92 = D(5)$ for $D(X) = 4X^2 - 8$. There are a lot of gaps. For $d = 19$, one must go to

$$D(X) = 9X^2 + 8X + 2, \quad P(X) = 81X^2 + 72X + 17, \quad Q(X) = 27X + 12,$$

choosing $m = 1$ for X. The most notorious value of d is $d = 61$. Bhāskara found its fundamental solution fairly easily. When the Europeans got involved, Fermat proposed solving

$$X^2 - 61Y^2 = 1$$

to Bernard Frénicle de Bessy (c. 1605 - 1675) in a letter written in 1657 shortly before sending his general challenge to the English mathematicians. The first to publish a solution was Euler in the next century. Some indication of how anomalous a value it is might be given by noting that the simplest reduced quadratic D representing 61 for which the Pellian (20) is solvable is

$$D(X) = 100X^2 + 44X + 5,$$

where $61 = D(-1)$, and the solution is $P(-1), Q(-1)$ for

$$P(X) = 7812500000X^6 + 10312500000X^5 + 5690625000X^4$$
$$+ 1680250000X^3 + 279975000X^2 + 24961200X + 930249,$$

$$Q(X) = 781250000X^5 + 859375000X^4 + 379375000X^3$$
$$+ 84012500X^2 + 9332500X + 416020.$$

I think I shall end our discussion of the Pell equation here, leaving as a parting exercise for the reader the verification that P, Q do indeed satisfy (20) for the given D and the calculation of the fundamental solution $P(-1), Q(-1)$ to (3) for $d = 61$. He/she might like to check this work by applying Bhāskara's algorithm to get an independent determination of this fundamental solution.

5

The Euclidean Algorithm

The question of Euclid's algorithm for finding the greatest common divisor of two integers came up in class. It is a fairly simple matter, but one of great importance.

The brute force method of finding the greatest common divisor of two numbers is to factor each of them completely and then take the product of the prime factors they have in common. For example, if we choose numbers 1000 and 275,

$$1000 = 10 \cdot 10 \cdot 10 = 2 \cdot 5 \cdot 2 \cdot 5 \cdot 2 \cdot 5 = 2^3 \cdot 5^3$$
$$275 = 5 \cdot 55 = 5 \cdot 5 \cdot 11 = 5^2 \cdot 11.$$

Their greatest common divisor is $5^2 = 25$ since the prime 5 divides each number twice.

For such small numbers, factorisation is easy and there is no need for a different method. For large numbers, however, factoring can be a time consuming task. For example, if we choose the numbers 10578 and 21285, we get

$$10578 = 2 \cdot 5289 = 2 \cdot 3 \cdot 1763 = 2 \cdot 3 \cdot 41 \cdot 43$$
$$21285 = 5 \cdot 4257 = 5 \cdot 9 \cdot 473 = 3^2 \cdot 5 \cdot 11 \cdot 43$$

and the greatest common divisor is $3 \cdot 43 = 129$.

The main point here is the factoring. For 1000, I noticed that $1000 = 10^3$ and took a short cut. Here, I noticed that 10578 is even so I divided by 2 right away, and that 21285 is divisible by 5 and I divided it by 5 right away. One can also use the trick that tells us that if the sum of the digits of a number is divisible by 3 or 9 then the number itself is divisible by 3 or 9, respectively. Thus, since $4 + 2 + 5 + 7 = (4 + 5) + (2 + 7) = 9 + 9$, I knew right away to divide by 9. For the most part, however, one is better off simply making a list of primes $2, 3, 5, \ldots$ and trying to divide by them in succession. Thus, for 10578 I tried 2, then 3, and then $5, 7, 11, 13, 17, 23, 29, 31, 37, 41$ in succession until I found the factor 41. Note that, if a number N is not prime, it has a prime factor $\leq \sqrt{N}$ [1] and $\sqrt{1763} \approx 41.9881$, so we had to do a maximal search to find a prime divisor (but we did not have to go as high as 101, the greatest prime less than $\sqrt{10578} \approx 102.85$). Note also that the primality of the quotient 43 has been checked: Any prime divisor of 43 would be a divisor of $41 \cdot 43 = 1763$ and we have already checked the candidates $2, 3, 5 < 6.557 \approx \sqrt{43}$.

The factorisation of 21285 was simpler as we only had to test for division by $2, 3, 5, 7, 11$ before completing the factorisation.

A slight change in numbers can make a big difference in the amount of labour involved, especially if one is computing by hand as was done in my student days. Consider

$$10579 = 71 \cdot 149$$
$$21287 = 7 \cdot 3041.$$

Factoring the former required me to try dividing 10579 by all primes $2, 3, 5, 7$, $11, 13, 17, 19, 23, 29, 31, 37, 41, 43, 47, 53, 59, 61, 67, 71$. Since $\sqrt{149} < 13 < 71$, I did not have to test it at all for prime factors as any prime divisor would be a prime divisor of 10579 and they had already been tested and ruled out. As for 21287, after dividing by 7, I had to test 3041 successively by the primes ≤ 53 before concluding 3041 to be prime. Needless to say I did all of this by programming my calculator to do all the testing for me. Anyway, once one has the factorisations, one sees there are no factors in common and the greatest common divisor of 10579 and 21287 is thus 1.

The key to finding the greatest common divisor of two numbers more efficiently is the Division Algorithm. If m, n are positive integers we may divide m by n to get a quotient q and remainder r with r smaller than n:

$$m = q \cdot n + r, \quad 0 \leq r < n.$$

Now any common divisor of m, n will also divide m, $q \cdot n$ and hence their difference r. Conversely, any divisor of n and r will divide $q \cdot n$ and r, hence their sum m. In symbols, writing $\gcd(x, y)$ for the greatest common divisor of x, y, we have

$$\gcd(m, n) = \gcd(n, r).$$

Or, writing $\operatorname{rem}(x, y)$ for the remainder of x after division by y,

$$\gcd(m, n) = \gcd(n, \operatorname{rem}(m, n)).$$

So to find the greatest common divisor of m, n we try to find the greatest common divisor of

$$m_1 = n, \quad n_1 = \operatorname{rem}(m, n)$$

and to find $\gcd(m_1, n_1)$, we look for $\gcd(m_2, n_2)$, where

$$m_2 = n_1, \quad n_2 = \operatorname{rem}(m_1, n_1).$$

---

[1] For, if N is composite, and $p \leq q$ are the smallest primes dividing N, then $p^2 \leq p \cdot q \leq N$, whence $p \leq \sqrt{N}$.

And so on until we can no longer perform the division, which happens when the n-value of the pair is 0. When that happens, the greatest common divisor is the current m-value.

We illustrate the procedure with our initial values 1000, 275:

$$1000 = 3 \cdot 275 + 175,$$

whence

$$\gcd(1000, 275) = \gcd(275, 175).$$

But

$$275 = 1 \cdot 175 + 100$$
$$175 = 1 \cdot 100 + 75$$
$$100 = 1 \cdot 75 + 25$$
$$75 = 3 \cdot 25 + 0,$$

i.e.,

$$\gcd(1000, 275) = \gcd(275, 175) = \gcd(175, 100)$$
$$= \gcd(100, 75) = \gcd(75, 25)$$
$$= \gcd(25, 0) = 25,$$

since 25 divides both 0 and 25.

This is not much of an improvement on factoring in this case. But now consider what happens with our larger numbers.

$$21285 = 2 \cdot 10578 + 129$$
$$10578 = 82 \cdot 129$$

and

$$\gcd(21285, 10578) = \gcd(10578, 129) = \gcd(129, 0) = 129.$$

Likewise

$$21287 = 2 \cdot 10579 + 129$$
$$10579 = 82 \cdot 129 + 1$$
$$129 = 129 \cdot 1 + 0$$

and $\gcd(21287, 10579) = 1$.

1 Exercise. Find the greatest common divisor of 10577 and 21284 using both the method of factoring and the Euclidean Algorithm. Repeat with the pair 10577, 21287.

The algorithm for finding the greatest common divisor is a favourite in beginning programming courses as an example of the use of *recursion*—allowing a program to call itself during execution. The gcd function is recursively defined by

$$\gcd(x,y) = \begin{cases} x, & \text{if } y = 0 \\ \gcd(y, \operatorname{rem}(x,y)), & \text{otherwise.} \end{cases}$$

In computer languages admitting recursion, this is almost already a program. In LOGO the program is written as follows:

```
TO GCD :X :Y
IF :Y=0 [OUTPUT :X]
OUTPUT GCD :Y REMAINDER :X :Y
END
```

On the *TI-83 Plus*, one can also define such a program, albeit not as simply. The calculator does not allow one to define functions, so one has to pass variables from the master program to the called program. Assuming values already stored in the variables X, Y, the program looks like this:

```
PROGRAM: GCD2
:If Y=0
:Then
:Disp X
:Else
:Y→A
:X−int(X/Y)∗Y→Y
:A→X
:prgmGCD2
:End
```

Of course, the calculator also has a function gcd(in the NUM submenu of the menu obtained by pressing the MATH button and one can quickly obtain the solutions to the above Exercise by punching in

gcd(10577,21284) and gcd(10577,21287),

respectively. The preprogrammed gcd(function will pretty much follow the same steps as the program. With the program, however, one can follow the progress of the algorithm by inserting a few extra command lines. On the calculator, one can place the commands

```
:Disp {X, Y}
:Pause
```

before the If statement. In execution, the program will display the current values of X, Y as a list and then pause to allow one to look at them before pressing the ENTER button and viewing the next pair. The Pause command is unnecessary, but the pairs tend to scroll rapidly and all but the last few will scroll off the window before one can read them. (In LOGO, one would place

```
SHOW (LIST :X :Y)
```

before the IF command. The computer screen allows many more lines than the calculator screen, so adding a pause is hardly necessary.)

The Euclidean Algorithm is so called because it is described by Euclid in *The Elements*, although it is probably a couple of centuries older. Euclid did not take a computational view of the algorithm, but a geometrical one. Following some preliminary definitions, he opened Book VII of *The Elements* with the algorithm—twice: Proposition VII-1 presents the algorithm in the case of relatively prime numbers, and Proposition VII-2 then handles the case of a proper common divisor.

Book VII begins Euclid's treatment of number theory and it is a bit strange. His definition of number is general, abstract, and not very precise; his actual presentation of the subject is geometric; and behind it all is lurking an intuitive conception of the counting numbers.

The first definition reads:

1. An *unit* is that by virtue of which each of the things that exist is called one.

I'm not sure what he means. The customary interpretation is that the unit is some arbitrarily chosen line segment to be used to measure other segments by, and numbers will be those segments that can be measured exactly by the unit. Formally, the definition of number goes:

2. A *number* is a multitude composed of units.

One often reads that in Greek mathematics, the unit generates numbers, but is not a number itself as "number" is to denote plurality. Whether Euclid takes this to be the case or not can be argued either way, at least if we use Heath's standard English translation. Definitions 3 and 22, combined with Proposition 36 of the subsequent Book IX suggest that he takes 1 to be a number. The two definitions read:

3. A number is a *part* of a number, the less of the greater, when it measures the greater.

22. A *perfect number* is that which is equal to its own parts.

A *part* is basically a divisor. A perfect number is one that equals the sum of its divisors other than itself. Euclid's Proposition IX-36 is the synthetic portion of his celebrated characterisation of even perfect numbers: If 2^{n-1} is prime, then $(2^n - 1) \cdot 2^{n-1}$ is perfect. In particular 6 is perfect. But $6 = 3 + 2 + 1$ and it will not be perfect unless we consider 1 to be a part—and the definition of "part" just given presupposes it to be a number.[2][3]

Against this let me cite Definitions 11 - 14:

---

[2] I suppose I should say somewhere that one can translate Euclid roughly as follows: part = divisor, measures = divides, common measure = common divisor, greatest common measure = greatest common divisor. Some inexactness, or confusion, arises in the cases where the would-be divisor is 1 or where we would not allow "2 numbers" to include the case where both numbers are the same.

[3] Note, by the way, that a part is strictly smaller than the number of which it is a part. This is stated explicitly in Definition 3 and is necessary in Definition 22 for the existence of perfect numbers: if a number were a part of itself, since it has 1 as a part, it would be less than its parts and there would be no perfect

11. A *prime number* is that which is measured by an unit alone.

12. Numbers *prime to one another* are those which are measured by an unit alone as a common measure.

13. A *composite number* is that which is measured by some number.

14. Numbers *composite to one another* are those which are measured by some number as a common measure.

To make sense, each of Definitions 13 and 14 requires the unit not to be a number. For, the unit measures all numbers and 13 would make all numbers composite if the unit, i.e. 1, were taken to be a number; and 14 would make any two numbers composite to one another. But Definitions 11 and 13, as well as 12 and 14, are clearly intended to define pairs of complementary concepts. Hence the unit is not a number even though it belongs with the numbers and is put to the same uses. In practical terms this means that Euclid must offer two statements of the basic result about the Euclidean Algorithm and two separate proofs, even though he was aware they had a common proof—which he gives later in Book X.

Euclid's algorithm is stated as Propositions 1 and 2 of Book VII:

2 Theorem (Proposition VII-1). *Two unequal numbers being set out, and the less being continually subtracted in turn from the greater, if the number which is left never measures the one before it until an unit is left, the original numbers will be prime to one another.*

3 Theorem (Proposition VII-2). *Given two numbers not prime to one another, to find their greatest common measure.*

As is occasionally the case in *The Elements*, the second proposition is not a proposition in the modern sense of a statement asserting a fact, but a statement asserting what he proposes to do. And he does this using a variant of the Euclidean Algorithm. The algorithm Euclid actually uses differs from what we call the Euclidean Algorithm in one minor respect. Instead of applying the Division Algorithm to express the larger of the two magnitudes as an integral multiple of the smaller magnitude plus an even smaller remainder in case the smaller magnitude does not measure the larger, he simply subtracts the smaller from the larger. The net result is that he has many more steps, but each uses simple subtraction and no division. Thus, if we return to our initial example, if the larger magnitude is 1000 and the smaller is 275, he would consider

$$\gcd(1000, 275) = \gcd(725, 275) = \gcd(450, 275)$$
$$= \gcd(275, 175) = \gcd(175, 100)$$
$$= \gcd(100, 75) = \gcd(75, 25)$$
$$= \gcd(50, 25) = \gcd(25, 25) = 25,$$

---

numbers. Likewise, Definition 11 tells us that numbers do not measure themselves. So I suppose for Euclid, the greatest common measure does not exist for a single number paired with itself!

which sequence has 4 more entries than the sequence given on page 73, above. Also, his last step is an equation $\gcd(x, x) = x$ and not $\gcd(x, 0) = x$ as before.

Proposition VII-1 explicitly assumes the algorithm to terminate in a unit and states as a conclusion that the given numbers are prime to one another. The statement of Proposition VII-2 is incomplete and one must read the proof to see what is established. One might expect the Proposition to be that, if two numbers are not prime to one another, this application of the algorithm terminates and the result is their greatest common measure. In fact, examination of the proof shows that termination is assumed. What Euclid does establish is that it must terminate in a number, not a unit (for, otherwise, Proposition VII-1 would make the given numbers prime to one another, contrary to the Proposition's assumption) and that the terminal number is the greatest common measure.

I would compare Propositions VII-1 and VII-2 to Euclid's Propositions III-20 and III-31. Proposition III-20, which we came across in discussing Ptolemy's Theorem in trigonometry, asserted that the angle at the centre of a circle determined by a segment was double the angle at the circumference determined by the same segment, and Propositions III-31 asserted the same for the special case in which the first angle was 180°. By his not considering the 180° angle to be an angle, this was a separate, not a special case to Euclid and it required a separate statement and proof. Similarly, because the unit is not a number, the statement that the algorithm terminates in the greatest common measure requires two statements and proofs for Euclid, although the necessity may elude us, who see the two results as two cases of a single result with a single proof.

Euclid's proofs of Propositions VII-1 and VII-2 would not pass muster today, logically because they lack a proof of termination and æsthetically because they needlessly split a common proof into two cases. But he does have the correct idea. If two numbers, represented by line segments AB and CD in Euclid, are not prime to one another, they have a number E serving as common measure. If one now subtracts the smaller, say CD from the larger AB, E must also measure the remainder. Likewise E measures the difference of the smaller segment and its remainder. And so on, until one is left with E measuring the last segment, which is assumed in the statement of Proposition VII-1 to be a unit. Thus, we have a contradiction and we conclude AB, CD to have no number serving as a common measure, i.e., AB and CD are prime to one another.

For Proposition VII-2, Euclid reasons that the final magnitude produced by the algorithm must be a number since otherwise VII-1 applies and the given numbers are prime to one another, contrary to assumption. He then reasons as before that the final number is a common measure and that it must itself be measured by any common measure and thus be the greatest common measure.

I suspect that the text is corrupted because at one point in the proof he writes, "If now CD measures AB—and it also measures itself—CD is a common measure of CD, AB"; yet Definition 11 makes clear that a prime number does not measure itself.

Perhaps Euclid has simply been sloppy because he had a better, more uniform proof to give in Book X. This book opens with a few definitions, most importantly the first:

> 1. Those magnitudes are said to be *commensurable* which are measured by the same measure, and those *incommensurable* which cannot have any common measure.

Following the definitions is the first proposition, a technical lemma that will prove useful in arguments involving limits, such as the proof in Book XII of his theorem on the areas of circles.

4 Lemma (Proposition X-1). *Two unequal magnitudes being set out, if from the greater there be subtracted a magnitude greater than its half, and from that which is left a magnitude greater than its half, and if this process be repeated continually, there will be left some magnitude which will be less than the lesser magnitude set out.*

In terms of real numbers, the lemma says that if x_0, ϵ are two positive real numbers, x_0 the larger, and if one defines a sequence $x_0, x_1, x_2, \ldots$ with $0 < x_{n+1} < \frac{1}{2}x_n$ for all n, then for some n one will have $x_n < \epsilon$. We would prove this by observing that

$$x_{n+1} < \frac{1}{2}\,x_n < \frac{1}{4}\,x_{n-1} < \ldots < \frac{1}{2^n}\,x_0,$$

and then choosing n so large that

$$\frac{1}{2^n}\,x_0 < \epsilon.$$

We do this by noting that, for $n > 0$,

$$\frac{1}{2^n} < \frac{1}{n},$$

and we have

$$\frac{1}{n}\,x_0 < \epsilon$$

for any $n > x_0/\epsilon$. That there is such an n is guaranteed by the Archimedean Axiom (Definition V-4) to which Euclid explicitly refers.[4]

Armed with this Lemma, he proves two important results about the Euclidean Algorithm.

5 Theorem (Proposition X-2). *If, when the less of two unequal magnitudes is continually subtracted in turn from the greater, that which is left never measures the one before it, the magnitudes will be incommensurable.*

---

[4] The Definition states that two magnitudes are capable of having a ratio if some multiple of each exceeds the other. He does not explicitly lay down the assumption as an axiom, but he uses it as such here and elsewhere in *The Elements*.

6 Theorem (Proposition X-3). *Given two commensurable magnitudes to find their greatest common measure.*

The proof of the first of these Propositions requires, in addition to the Lemma, another application of the Archimedian Axiom to conclude the Division Algorithm: Given two magnitudes AB and CD, with AB greater than CD and CD not measuring AB, one can subtract some multiple of CD from AB to obtain a remainder AE less than CD. Now AE is less than CD and $CD + AE$ is less than or equal to AB, whence AE is less than half of AB.

The rest is now easy. Let AB, CD be commensurable with common measure E. If we can keep applying the Division Algorithm, we eventually obtain a segment smaller than E. But if E measures AB and CD, it measures the difference and the next difference and the next... So it measures a magnitude smaller than itself, a contradiction.

The derivation of Proposition X-3 from Proposition X-2 should now be routine and I leave it to the reader.

With the proof of Proposition X-3, an adequate common proof of Propositions VII-1 and VII-2 has been given. The relation between these results and the numbers of Book VII is clarified in a series of Propositions (X-5 to X-8) which combine to assert that two magnitudes are commensurable iff their ratio is the ratio of two numbers. We would express this more simply by saying their ratio is a *rational* number, but Euclid would not do so because i. "number" to him meant whole number, and ii. he used the word "rational" in a different sense in Book X, a matter I do not want to go into here.

What I do wish to discuss is the application of Proposition X-2, i.e., the application of the Euclidean Algorithm, to irrational numbers. In practice, applying the algorithm to a pair of irrational numbers can be a computational morass that quickly overwhelms one's calculator—and computer as well if one isn't careful. However, with quadratic irrationals it is workable.

The easiest example is given by the *golden ratio*, denoted ϕ. This is the ratio of the length to width of a rectangle with the curious reproductive property that, if one subtracts the shorter width from the length and creates a new rectangle using the width and shortened length, respectively, the ratio of length

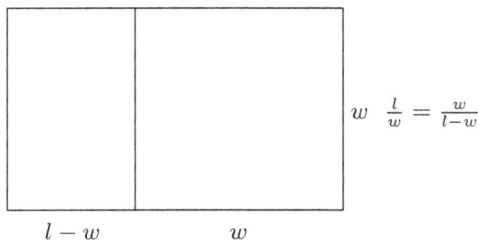

$$w \quad \frac{l}{w} = \frac{w}{l-w}$$

$l - w$ ㅤㅤ w

FIGURE 1.

to width remains the same. (See FIGURE 1.) If we set the original length to ϕ and the width to 1, this reads

$$\frac{\phi}{1} = \frac{1}{\phi - 1},$$

i.e.,

$$\phi^2 - \phi = 1,$$

which equation has the solutions

$$\frac{1 \pm \sqrt{5}}{2},$$

yielding

$$\phi = \frac{1 + \sqrt{5}}{2},$$

since we take the length to be positive.

We do not need to do any actual calculation to demonstrate that application of the Euclidean Algorithm to $\phi, 1$ never terminates. We note that the length and width of the first rectangle is in the ratio ϕ which is greater than 1, so we can subtract 1 from ϕ. The dimensions of the next rectangle are in the same ratio, so we can subtract its width from its length. But the resulting rectangle has the same dimension and we can subtract its width from its length to obtain yet the same ratio. The process clearly never stops.

The earliest numbers cited as irrational in the extant literature are $\sqrt{3}$, $\sqrt{5}$, $\sqrt{6}$, $\sqrt{7}$, $\sqrt{8}$, $\sqrt{10}$, $\sqrt{11}$, $\sqrt{12}$, $\sqrt{13}$, $\sqrt{14}$, $\sqrt{15}$, $\sqrt{17}$, which were mentioned by Plato (427 - 348/347 B.C.) in his dialogue *Theætetus*[5] . The irrationality of $\sqrt{2}$ is presumed already to have been known and the irrationalities of the numbers cited an announcement of recent results of Theodorus (*c.* 465 - *after* 399 B.C.). The irrationality of $\sqrt{2}$ is mentioned with a passing reference to its proof of irrationality by Aristotle. However, the naturalness of the above proof of the irrationality of ϕ, when combined with the Pythagorean adoption of the *pentagram* (the 5-pointed star inscribed in the regular pentagon) as a special symbol and the numerous appearances of ϕ in the pentagram, have led to the conjecture that ϕ was indeed the first number proven irrational by the Pythagoreans.

---

[5] Named after Theætetus (*c.* 417 - 364 B.C.), who is credited with much of the theory of irrationals found in Book X of *The Elements*, much of Book XIII of *The Elements* on the five regular solids, and is considered a probable precursor to Eudoxus in establishing a general theory of proportion applicable to both commensurable and incommensurable magnitudes.

6

The Fibonacci Numbers

The famous mathematician Leonardo of Pisa (*c.* 1170 - *c.* 1240) was a member of an Italian family named Bonacci. The Italian word for "son" being *figlio* (from the Latin *filius*), he came to be called Fibonacci, or son of the Bonaccis. His great contribution to mathematics and Europe in general was his book *Liber abbaci* (1202), the book of calculation[1], which was the first widely accessible European description of the Hindu-Arabic numerals. It also contained many word problems, both practical and theoretical. One of these is his famous rabbit problem:

> How many pairs of rabbits can be bred from one pair in one year?
> A man has one pair of rabbits at a certain place entirely surrounded by a wall. We wish to know how many pairs can be bred from it in one year, if the nature of these rabbits is such that they breed every month one other pair and begin to breed the second month after birth.[2]

---

[1] One finds the book variously cited as *Liber abaci* and *Liber abbaci*. I follow the *Dictionary of Scientific Biography* on this point.

[2] Translation from Dirk Struik (ed.), *A Source Book in Mathematics, 1200 - 1800*, Harvard University Press, Cambridge (Mass), 1969, pp. 2 - 3. Struik includes the full text of the problem, including Fibonacci's solution. The statement of the problem is repeated in other sources with minor differences in translation. Victor J. Katz, *A History of Mathematics; An Introduction*, HarperCollins College Publishers, 1993, p. 284, offers what appears to be a more literal translation, while Ian Stewart, *The Story of Mathematics; From Babylonian Numerals to Chaos Theory*, Quercus Publishing Plc, London, 2008, p. 61, polishes the English of his translation. Against this I cite the translation given in Sam E. Ganis in the capsule on the Fibonacci numbers in *Historical Topics for the Mathematics Classroom*, National Council of Teachers of Mathematics, Washington, D.C., 1969, p. 77:
> What is the number of pairs of rabbits at the beginning of each month if a single pair of newly born rabbits is put into an enclosure at the beginning of January and if each pair breeds a new pair at the beginning of the second month following birth and an additional pair at the beginning of each month thereafter?

Not only the wording, but the problem itself is different. It is now explicitly assumed that the original pair is newly born, and one is asked not for a final result, but also

It is, in fact, this problem and the amazing sequence of numbers that grew out of its solution for which Fibonacci is most remembered today.

Fibonacci reasoned simply as follows: One starts with one pair of rabbits. At the end of the first month there is a new pair, whence 2 in all. At the end of the next month, the first pair gives birth to a new pair but the second pair is not yet old enough, so only one pair has been added to the 2 previous, leaving one with $2 + 1 = 3$ pairs of rabbits. Going into the next month, one has 3 pairs of rabbits, 2 of which are of breeding age, whence the month will finish off with $3 + 2 = 5$ pairs. More generally, one has the sequence $1, 2, 3, 5, 8, 13, 21, 34, 55, 89, 144, 233, 377, \ldots$, where each entry after the second is the sum of the two previous entries—the number of pairs (the previous entry) and the number of breeding pairs (the entry just prior to the previous entry). And counting to the end of the 12th month, we see that the answer is 377 (i.e., the 13th entry).

Today we extend the sequence backwards 2 steps and call the sequence,

$$0, 1, 1, 2, 3, 5, 8, \ldots,$$

the *Fibonacci sequence*, or, more loosely, the *Fibonacci numbers* and denote them by the letter f, thus

$$f_0, f_1, f_2, \ldots$$

or

$$F_0, F_1, F_2, \ldots$$

My own preference is for the lower case, but one can find both in the literature.

The modern approach is only to pay lip service to Fibonacci's rabbit problem and emphasise the infinite sequence. To this end, one has the formal definition:

1 Definition. *The* Fibonacci sequence *is the infinite sequence* $f_0, f_1, f_2, \ldots$ *of numbers defined by the following recursion:*

$$f_0 = 0$$
$$f_1 = 1$$
$$f_{n+2} = f_n + f_{n+1}.$$

The word "recursion", as mentioned in our discussion of the Euclidean Algorithm in the preceding chapter, is used in computer science to describe programs that call themselves as subroutines during execution. The traditional mathematical term for the "recursive call" is a *recurrence relation*. Recurrent sequences are defined by specifying k initial elements and a recurrence

for a running tally as it were. Not surprisingly, Ganis comes up with a different final result than does Leonardo in Struik's translation. The *Liber abbaci* was revised in 1228, so possibly Ganis quotes a different version of the problem from the others. Or he simply got it wrong. Curiously, the references cited by him include Struik's book, yet he does not mention the difference.

relation specifying how each later element is defined in terms of its k immediate predecessors. The Fibonacci sequence is an example of a sequence defined by a *linear recurrence*, i.e., one in which a_{n+k} is a linear combination of $a_n, a_{n+1}, \ldots, a_{n+k-1}$: for some $b_0, \ldots, b_{k-1}$, one has

$$a_{n+k} = b_0 a_n + \ldots + b_{k-1} a_{n+k-1}$$

for all n. Such recurrences are the easiest to analyse and often have interesting properties, a state of affairs that made such sequences, especially under the 19th century French mathematician Édouard Anatole Lucas (1842 - 1891), a centrepiece of recreational mathematics. Of particular interest to the recreational mathematician is the sequence of Fibonacci numbers and, indeed, it is probably the most studied sequence after the sequences of natural numbers and prime numbers. Before considering anything serious, it might be amusing to take a quick look at the recreational aspects of the Fibonacci numbers.

We might start by making a small table of values of f_n, f_{n+1} as in TABLE 1, and examining it.

n	0 1 2 3 4 5 6 7 8 9 10 11 12
f_n	0 1 1 2 3 5 8 13 21 34 55 89 144
f_{n+1}	1 1 2 3 5 8 13 21 34 55 89 144 233

TABLE 1.

Comparing the entries in the second and third rows, we might notice that the entries in a given column, i.e., f_n and f_{n+1}, have no factor in common. Or we might notice that the even numbers in the second row are $f_0, f_3, f_6, f_9, f_{12}$, those divisible by 3 are f_0, f_4, f_8, f_{12}, those by 5 are f_0, f_5, f_{10}. In fact, we might make up another table (TABLE 2) listing those Fibonacci numbers divisible by a given divisor. Can you guess the first f_n after f_0 that is divisible by 10? By

n	f_k's divisible by n
2	$f_0, f_3, f_6, f_9, f_{12} \ldots$
3	$f_0, f_4, f_8, f_{12}, f_{16}, \ldots$
4	$f_0, f_6, f_{12}, f_{18}, f_{24}, \ldots$
5	$f_0, f_5, f_{10}, f_{15}, f_{20}, \ldots$
6	$f_0, f_{12}, f_{24}, f_{36}, f_{48}, \ldots$
7	$f_0, f_8, f_{16}, f_{24}, f_{32}, \ldots$
8	$f_0, f_6, f_{12}, f_{18}, f_{24}, \ldots$
9	$f_0, f_{12}, f_{24}, f_{36}, f_{48}, \ldots$

TABLE 2.

42?

One might decide to make a table of squares of Fibonacci numbers (TABLE 3) and compare f_n^2 with f_{n+1}^2, perhaps adding the two values. It shouldn't

n	0 1 2 3 4 5 6
f_n^2	0 1 1 4 9 25 64
f_{n+1}^2	1 1 4 9 25 64 169
$f_n^2 + f_{n+1}^2$	1 2 5 13 34 89 233

TABLE 3.

take one too long to recognise the elements of the bottom row and formulate a conjecture:

$$f_n^2 + f_{n+1}^2 = f_{2n+1}.$$

The Fibonacci numbers have so many nice properties that it would seem anything one tries will result in something interesting. For example, one might try adding the first $n+1$ Fibonacci numbers as in TABLE 4. Do you see an

n	0 1 2 3 4 5 6 7
f_n	0 1 1 2 3 5 8 13
sum to f_n	0 1 2 4 7 12 20 33

TABLE 4.

interesting result? If not, add 1 to each of the entries in the bottom row and you will surely recognise it.

Discovering such relations is half the fun. The other half is proving their correctness.

The Fibonacci sequence is a recurrent sequence, its successive elements being defined inductively. The natural method of proving results about them would thus be mathematical induction.

Mathematical induction has a murky history, with a growing list of contenders for the honour of having been its author: the Pythagoreans, Plato, Euclid, Abū Kāmil Shujā' Ibn Aslam (c. 850 - c. 930), al-Karajī (fl. c. 1000), al-Samaw'al (died c. 1180), Rabbi Levi ben Gerson (1288 - 1344), Francesco Maurolico (1494 - 1575), Blaise Pascal (1623 - 1662), and Jakob Bernoulli. This list covers an interval of over two millenia. Why so long? The answer is supplied by a view put forth in various popular works by Henri Poincaré (1854 - 1912) in the early 20th century: mathematical induction is a special mathematical intuition.

The intuitive nature of induction explains the long list of candidates for the designation as the principle's originator. The principle has been used since the early days of scientific mathematics more-or-less as a law of thought, without mention at first, and then only gradually being rendered more explicit until finally Pascal gave it almost formal expression in referring to the proof of some proposition:

> Although this proposition has an infinite number of cases, I shall give a very short demonstration of it based on 2 lemmata. The first, which is self-evident, is that this proportion holds in the second base...
>
> The second is that if this proposition holds in some arbitrary base then it necessarily also holds in the following base.
>
> Whence we see that it is necessary in all the bases: for if it is in the second base, by the first lemma, therefore by the second it is in the third base, therefore in the fourth, and so on to infinity.[3]

This is not as explicit as, say,

> If the number 1 has a certain property, and if, whenever n has that property, so does the (next) number $n+1$, then every number possesses the property.[4]

But this is a matter of exposition, not understanding. Pascal has clearly described, albeit in a specific case, the two steps of an inductive proof—the basic step showing something holds for the first number (in his case 2) and the inductive step showing its truth carries over from one number to the next. Bernoulli would later (1686) describe the procedure in another special instance and go on to state explicitly that the method is quite general.

Today, we would probably start our natural numbers at 0 and state the method rather formally:

> Given a property P of numbers[5], to prove that every number n has property P,
>
> $$\text{for all } n, \ P(n) \text{ holds,}$$
>
> it suffices to prove $P(0)$ holds and
>
> $$\text{for all } n, \text{ if } P(n) \text{ holds then } P(n+1) \text{ holds.}$$

Indeed, one might go very formal and state the principle as an *inference rule*:

> from $P(0)$ and $\forall n \big(P(n) \to P(n+1) \big)$
> conclude $\forall n \, P(n)$,

---

[3] I quote the English translation of this passage from Pascal's treatise (1665) on what we now call Pascal's triangle, the *Traité du triangle arithmétique* given in John Crossley, *The Emergence of Number*, World Scientific, Singapore, 1987, p. 44. The second chapter of Crossley's book outlines the gradual development of the formal procedure of proof by induction.

[4] *Ibid.*, p. 3.

[5] P may also depend on other parameters and we could write $P(n,m)$ instead of $P(n)$. I suppress the "m" below to emphasise the variable of the induction.

where "$\forall n$" reads "for all natural numbers n" and "$A \to B$" reads "if A then B" or "A implies B". One advantage of this formulation is that it makes clear that the variable n is a *bound* variable and can be replaced by any other, e.g., k. That is,

> from $P(0)$ and $\forall k\big(P(k) \to P(k+1)\big)$
> conclude $\forall n\, P(n)$,

The standard proof by induction proceeds as follows:

Theorem. For all n, $P(n)$.
Proof. Basis Step $(n = 0)$. [$P(0)$ is proven.]
Induction Step. Assume $P(n)$. [A derivation of $P(n+1)$ from the assumption is given.] Therefore $P(n+1)$.
Conclusion. $P(n)$.

This proof can be quite confusing for beginners. I am inclined to believe the difficulty is with the formalism, not the conception. For, students always understand the domino metaphor.[6] The difficulty is with the formalism. There are, of course, the standard difficulties students have with formalisms, but here there is also the confusion wrought by assuming $P(n)$—the very assertion one is trying to prove—in the induction step. This latter confusion, however, can probably be cleared up simply by deriving $P(k+1)$ from $P(k)$ instead of $P(n+1)$ from $P(n)$.

The simplest proofs by induction are of equations. The standard introductory examples deal with sums of integers:

2 Theorem. *For any $n \geq 0$,*

$$0 + 1 + \ldots + n = \frac{n(n+1)}{2}.$$

Proof. Let $P(n)$ be the equation

$$0 + 1 + \ldots + n = \frac{n(n+1)}{2}.$$

We prove $\forall n\, P(n)$ by induction on n.
Basis $(n = 0)$. Obviously

$$0 = \frac{0(0+1)}{2}.$$

Induction Step. Assume $P(k)$:

$$0 + 1 + \ldots + k = \frac{k(k+1)}{2},$$

---

[6] If you stand up a bunch of dominoes in such a way that knocking one over will knock the next one down, then knocking over the first will result in all the dominoes falling. This example is so readily understood that it escaped the confines of the mathematics classroom and became the much hated "domino theory" of the Vietnam War era.

and add $k + 1$ to both sides:

$$0 + 1 + \ldots + k + (k + 1) = \frac{k(k + 1)}{2} + (k + 1)$$
$$= \frac{k(k + 1)}{2} + \frac{2(k + 1)}{2}$$
$$= \frac{k(k + 1) + 2(k + 1)}{2}$$
$$= \frac{(k + 2)(k + 1)}{2} = \frac{(k + 1)(k + 2)}{2},$$

i.e., $P(k + 1)$.

Conclusion. $P(n)$ holds by induction. $\qquad\square$

3 Exercise. Prove the following formulæ by induction:

$$0^2 + 1^2 + \ldots + n^2 = \frac{n(n + 1)(2n + 1)}{6}$$

$$0^3 + 1^2 + \ldots + n^3 = (0 + 1 + \ldots + n)^2.$$

The Fibonacci numbers and their numerous properties offer a good stock of results to practise proofs by induction on. For example, there is the identity cited earlier and first proven by Lucas:

4 Theorem. *For all $n \geq 0$,*

$$f_0 + f_1 + \ldots + f_n = f_{n+2} - 1.$$

Proof. By induction on

$$P(n): \quad f_0 + f_1 + \ldots + f_n = f_{n+2} - 1.$$

Basis ($n = 0$). $f_0 = 0 = 1 - 1 = f_2 - 1$.
Induction Step. Assume

$$P(k): \quad f_0 + f_1 + \ldots + f_k = f_{k+2} - 1,$$

and add f_{k+1} to both sides of the equation:

$$f_0 + f_1 + \ldots + f_k + f_{k+1} = f_{k+2} - 1 + f_{k+1}$$
$$= f_{k+2} + f_{k+1} - 1$$
$$= f_{k+3} - 1 = f_{(k+1)+2} - 1,$$

and we have $P(k + 1)$.

Concluson. $P(n)$ holds for all $n \geq 0$. $\qquad\square$

The property being inducted on does not have to be expressed by an equation.

5 Lemma. *For any $n \geq 1$, f_n and f_{n+1} are relatively prime.*

Proof. By induction on

$$P(n): \quad f_n \text{ and } f_{n+1} \text{ are relatively prime.}$$

Basis $(n = 1)$. $f_1 = f_2 = 1$, whence f_1, f_2 have only 1 as common divisor, i.e., f_1, f_2 are relatively prime.

Induction Step. Assume $P(k)$. Observe

$$f_{k+1} = f_k + f_{k-1}$$
$$f_{k+2} = f_{k+1} + f_k = (f_k + f_{k-1}) + f_k = 2f_k + f_{k-1}.$$

If d divides f_{k+1} and f_{k+2}, then it divides their difference,

$$(2f_k + f_{k-1}) - (f_k + f_{k-1}) = f_k.$$

That is, d divides f_{k+1} and f_k. But by $P(k)$, however, d must therefore be 1. Thus we have $P(k+1)$.

Conclusion. $P(n)$ holds for all $n \geq 1$. □

The choice of P may not always be so obvious.

6 Theorem (Fibonacci Addition Formula). *For* $m \geq 1$,

$$f_{n+m} = f_m f_{n+1} + f_{m-1} f_n.$$

What is P? Is the induction on m or n? Assuming m fixed and

$$P(n): \quad f_{n+m} = f_m f_{n+1} + f_{m-1} f_n,$$

the proof of the induction step goes

$$
\begin{aligned}
f_{(k+1)+m} = f_{k+m+1} = f_{k+m} + f_{k+m-1} &= f_{k+m} + f_{(k-1)+m} \qquad (1) \\
&= (f_m f_{k+1} + f_{m-1} f_k) + (f_m f_k + f_{m-1} f_{k-1}) \qquad (2) \\
&= f_m(f_{k+1} + f_k) + f_{m-1}(f_k + f_{k-1}) \\
&= f_m f_{k+2} + f_{m-1} f_{k+1},
\end{aligned}
$$

i.e., $P(k+1)$. The step from (1) to (2), however, is problematic.

First, there is the problem of what to do if $k = 0$. Notice that (2) has a term f_{k-1}, which would be f_{-1} for $k = 0$, and we have not defined f_{-1}. This is not much of a problem: We simply prove $P(1)$ directly without relying on $P(0)$ and reserve the induction step for $k \geq 1$. To this end, note

$$f_{1+m} = f_m + f_{m-1} = f_m f_2 + f_{m-1} f_1,$$

since $f_2 = f_1 = 1$.

Another problem is that we have concluded (2) not from $P(k)$, but from $P(k)$ and $P(k-1)$. Thus, our induction should be on

$$Q(n): \quad P(n) \ \& \ P(n-1),$$

and not on $P(n)$. If one doesn't like to induct on anything as complicated as Q, one can induct directly on P by using an auxiliary induction principle:

from $P(0), P(1)$ and $\forall k\big(P(k)\ \&\ P(k+1) \to P(k+2)\big)$
conclude $\forall n\, P(n)$.

The same Poincaréan special mathematical intuition that makes ordinary induction obvious to us should apply here also.

If we decide to do the induction on m, the induction step for

$$P(m): \quad f_{n+m} = f_m f_{n+1} + f_{m-1} f_n,$$

goes

$$f_{n+(k+1)} = f_{(n+1)+k} \tag{3}$$
$$= f_k f_{n+2} + f_{k-1} f_{n+1} \tag{4}$$
$$= f_k(f_{n+1} + f_n) + f_{k-1} f_{n+1}$$
$$= f_k f_{n+1} + f_k f_n + f_{k-1} f_{n+1}$$
$$= (f_k + f_{k-1}) f_{n+1} + f_k f_n$$
$$= f_{k+1} f_{n+1} + f_k f_n,$$

i.e., $P(k+1)$. Once again our first inference is problematic. For we have not concluded the equation (4) from (3) via $P(k)$, but via a similar result with n replaced by $n+1$. Our induction should actually be on $\forall n\, P(m)$, i.e.,

$$Q(m): \quad \forall n\, \big(f_{n+m} = f_m f_{n+1} + f_{m-1} f_n\big),$$

and not merely on $P(m)$. Then we can replace n in $P(k)$ by any value we choose in the given inference.

I leave to the reader the exercise of putting together a correct formal proof of Theorem 6.

The theorem has a couple of nice corollaries.

7 Corollary. *For all n, $f_n^2 + f_{n+1}^2 = f_{2n+1}$.*

Proof. Let $m = n+1$ in Theorem 6. $\qquad\square$

8 Corollary. *For all m, n, f_n divides f_{mn}.*

Proof. Since $m0 = 0$, the result is trivial for $n = 0$. Thus we assume $n > 0$. We prove the result for $n > 0$ by induction on

$$P(m): \quad f_n \text{ divides } f_{mn}.$$

Basis $(m = 0)$. f_n divides $0 = f_m$.
Induction Step. Assume $P(k)$. For $k = 0$, $(k+1)n = n$ and the result is trivial. Thus assume $k > 0$ and observe

$$f_{(k+1)n} = f_{kn+n} = f_{kn} f_{n+1} + f_{kn-1} f_n. \tag{5}$$

But f_n divides $f_{kn-1} f_n$ and, by $P(k)$, it also divides f_{kn}, whence f_n divides $f_{kn} f_{n+1}$. Thus f_n divides both terms on the right in (5) and therefore divides their sum $f_{(k+1)n}$. $\qquad\square$

[At some point in one's development one stops adding the concluding statement at the end of every proof by induction.]

And now for a real treat! In 1876, Édouard Lucas proved the following:

9 Theorem. *For all m, n, $f_{\gcd(m,n)} = \gcd(f_m, f_n)$.*

This is a very nice result. Corollary 8 can be invoked to partially explain the regularity of TABLE 2. If a number n divides f_m, then it also divides $f_{2m}, f_{3m}, f_{4m}, \ldots$ But in that table there were no extraneous insertions. In the row for $n = 3$, for example, f_4 is the smallest Fibonacci number divisible by 3, whence the list of divisors includes $f_0, f_4, f_8, \ldots$ But it includes no others. And Theorem 9 tells us why: If f_k were on the list, 3 would divide f_4 and f_k, whence it would divide $\gcd(f_4, f_k) = f_{\gcd(4,k)}$. Because f_4 is the first Fibonacci number divisible by 3, we have $4 \leq \gcd(4, k)$. But clearly $\gcd(4, k) \leq 4$. Thus $\gcd(4, k) = 4$ and 4 divides k.

As for the proof, we are once again in the situation of the proof of Theorem 6, where in inducting on n we concluded the truth of $P(k + 1)$ not from that of $P(k)$ but from the truths of $P(k)$ and $P(k - 1)$. In the proof to follow, we will conclude the truth of $P(k+1)$ from that of *some* assertion $P(i)$ for $i \leq k$. This situation occurs quite often in mathematics and it is not always possible to determine explicitly the number i as a function of k (or, of k and whatever other parameters might be lurking about), so one assumes as induction hypothesis $P(0), P(1), \ldots, P(k)$:

from $P(0)$ and the implication, for all k,

$$P(0) \ \& \ P(1) \ \& \ \ldots \ \& \ P(k) \rightarrow P(k + 1),$$

conclude $\forall n \, P(n)$.

Letting "$\forall i < k$" abbreviate "for all natural numbers $i < k$", we can rewrite this as

$$\left. \begin{array}{l} \text{from } P(0) \text{ and } \forall k \big(\forall i < k + 1 \, P(i) \ \rightarrow \ P(k+1) \big) \\ \text{conclude } \forall n \, P(n) \end{array} \right\}. \tag{6}$$

Noting that, there being no natural numbers less than 0, the assertions $P(0)$ and $\forall i < 0 \, P(i) \rightarrow P(0)$ are equivalent, one often sees (6) written more simply as

$$\left. \begin{array}{l} \text{from } \forall k \big(\forall i < k \, P(i) \ \rightarrow \ P(k) \big) \\ \text{conclude } \forall n \, P(n) \end{array} \right\}. \tag{7}$$

In practice, one often proves the premise to (7) in two cases: $k = 0$ and $k > 0$, and thus one is really using (6). But (7) is a little more elegant. Whichever way one formulates it, the principle is singled out from other variants of mathematical induction and is called the *Strong Form of Mathematical Induction*, or, in Germany, the *Principle of Complete Induction*, because it uses a weaker-looking hypothesis. It is equivalent to ordinary induction.

[The equivalence proof is easy once one sees the substitution for P in establishing the strong form from the ordinary one.

[Assume the strong form of induction (6) and suppose one has proven $P(0)$ and $P(k) \rightarrow P(k+1)$. From the assumption $\forall i < k + 1 \, P(i)$ we conclude $P(k)$ and thence $P(k+1)$:

$$\forall i < k+1 \, P(i) \; \rightarrow \; P(k+1).$$

Hence (6) applies and we conclude $\forall n \, P(n)$.

[The converse is tricker. We do not conclude (6) from the instance of ordinary induction on P, but from the principle's application to Q:

$$Q(n): \quad \forall i < n+1 \, P(i).$$

Now assume $P(0)$ and $\forall i < k+1 \, P(i) \; \rightarrow \; P(k+1)$. $Q(0)$ is just $\forall i < 1 \, P(i)$, i.e., $P(0)$, whence we have $Q(0)$.

[Now

$$
\begin{aligned}
Q(k) &\rightarrow \forall i < k+1 \, P(i), \text{ by choice of } Q\\
&\rightarrow P(k+1), \text{ by assumption}\\
&\rightarrow \forall i < k+1 \, P(i) \; \& \; P(k+1), \text{ combining the above two lines}\\
&\rightarrow \forall i < k+2 \, P(i)\\
&\rightarrow Q(k+1).
\end{aligned}
$$

We have now established $Q(0)$ and $\forall k \big(Q(k) \rightarrow Q(k+1) \big)$, whence ordinary induction yields $Q(n)$, i.e., $\forall i < n+1 \, P(i)$. In particular we have $P(n)$. As n was arbitrary, we conclude $\forall n \, P(n)$.]

Using the Strong Form of Mathematical Induction, we can quickly prove the Theorem:

Proof of Theorem 9. Since $\gcd(f_m, f_n) = \gcd(f_n, f_m)$ and $\gcd(m, n) = \gcd(n, m)$, we may assume $m \geq n$.

If $m = n$, then $f_m = f_n$ and

$$\gcd(f_m, f_n) = \gcd(f_m, f_m) = f_m = f_{\gcd(m,m)} = f_{\gcd(m,n)}.$$

Thus assume $m > n$. We prove the result by applying the strong form of induction to

$$P(m): \quad \forall n < m \big(\gcd(m, n) = \gcd(f_m, f_n) \big).$$

Basis $(m = 1)$. The only value of n less than 1 is $n = 0$:

$$\gcd(1, 0) = 1, \quad \gcd(f_1, f_0) = \gcd(1, 0) = 1 = f_1.$$

Induction Step. To avoid too much confusion with the relabelling of variables, I shall not make the customary substitution of k for m in the induction step. Thus assume $\forall i < m+1 \, P(i)$. Since $m > n$, we can write

$$f_m = f_{(m-n)+n} = f_{m-n}f_{n+1} + f_{m-n-1}f_n.$$

Now f_n, f_{n+1} are relatively prime by Lemma 5 and we see that any common divisor of f_m, f_n must divide f_{m-n}, and conversely any common divisor of f_{m-n}, f_n must divide f_m. Thus

$$\gcd(f_m, f_n) = \gcd(f_{m-n}, f_n).$$

Now, each of $m - n, n$ is less than m and, so long as they are not equal, we can apply the induction hypothesis to conclude

$$\gcd(f_m, f_n) = \gcd(f_{m-n}, f_n) = f_{\gcd(m-n,n)}.$$

However, it is easy to see that $\gcd(m - n, n) = \gcd(m, n)$.

In the case where $m - n = n$, we argue differently: $m = 2n$ and Corollary 8 tells us that f_n divides f_m, whence $\gcd(f_m, f_n) = f_n = f_{\gcd(m,n)}$. □

I don't suppose anyone would consider the proof of Theorem 9 to be much of a treat, but the Theorem itself is quite sweet. I should probably complement it with a proof that every number divides some Fibonacci number, but I think I shall leave that and other recreational explorations to the interested reader. There is a vast literature on the Fibonacci numbers, as a visit to the Internet will quickly reveal. One book on the Fibonacci numbers that has been popular for several decades and been republished several times is N.N. Vorobyov[7], *Fibonacci Numbers*. The current English edition is published by Birkhäuser, Boston.

The Fibonacci numbers also have their serious side outside recreational mathematics, and I should cite an example or two. First, let me cite them as a canonical example of how to write an inefficient program in computer science. The defining recursion translates easily into programs in some languages. For example, in LOGO one has

```
TO FIB :N
IF (:N=0) [OUTPUT 0]
IF (:N=1) [OUTPUT 1]
OUTPUT (FIB (:N - 1)) + (FIB (:N - 2))
END
```

This will work, but it is highly inefficient. Think about how the calculation goes. To calculate f_5, after determining 5 is neither 0 nor 1, it sets about to calculate f_4 and f_3. It checks that 4 is neither 0 nor 1 and then sets about calculating f_3 and f_2. Notice that f_3 is thus calculated twice. f_2 is calculated on its own, and once each as subcalculations in the two calculations of f_3. The duplication of effort is massive. It is amusing to count how many times each f_k is calculated as a subcalculation of f_n. To this end, define $c(n, k)$ for $n \geq k$ to be the number of calls to FIB k made in a calculation of FIB n. For $k = 0$,

$c(0, 0) = 0$, since FIB 0 outputs 0 making no calls

$c(1, 0) = 0$, since FIB 1 outputs 1 making no calls

$c(2, 0) = 1$, since FIB 2 calls FIB 1 and FIB 0, but FIB 1 is calculated directly

$c(3, 0) = 1$, since FIB 3 calls FIB 2 and FIB 1, FIB 1 makes no calls to anything and FIB 2 calls FIB 0 once.

After that
$$c(2 + n + 2, 0) = c(2 + n + 1, 0) + c(2 + n, 0).$$

___
[7] Other transliterations include Vorobiev and Vorobev.

So the sequence $c(0,0), c(1,0), c(2,0), \ldots$ is $0, 0, 1, 1, 2, 3, 5, \ldots$, i.e.,

$$c(n+1, 0) = f_n.$$

But f_n can be quite large:

10 Theorem. *For $n > 2$,*

$$\left(\frac{3}{2}\right)^{n-2} < f_n \leq 2^{n-2}.$$

Proof. By induction.
Basis $(n = 3, 4)$. We have

$$\left(\frac{3}{2}\right)^{3-2} = \frac{3}{2} < 2 = f_3 = 2^{3-2}$$

$$\left(\frac{3}{2}\right)^{4-2} = \frac{9}{4} < 3 = f_4 < 2^2 = 2^{4-2}.$$

Induction Step. Assume

$$\left(\frac{3}{2}\right)^{k-2} < f_k \leq 2^{k-2}$$

$$\left(\frac{3}{2}\right)^{k+1-2} < f_{k+1} \leq 2^{k+1-2}.$$

Then observe

$$\left(\frac{3}{2}\right)^{k-2} + \left(\frac{3}{2}\right)^{k+1-2} < f_k + f_{k+1} \leq 2^{k-2} + 2^{k-1}.$$

But

$$\left(\frac{3}{2}\right)^{k-2}\left(1 + \frac{3}{2}\right) = \left(\frac{3}{2}\right)^{k-2}\left(\frac{5}{2}\right) = \left(\frac{3}{2}\right)^{k-2}\left(\frac{10}{4}\right)$$

$$> \left(\frac{3}{2}\right)^{k-2}\left(\frac{9}{4}\right) = \left(\frac{3}{2}\right)^{k-2+2},$$

whence

$$\left(\frac{3}{2}\right)^{(k+2)-2} < f_k + f_{k+1} = f_{k+2}.$$

Similarly,

$$2^{k-2} + 2^{k-1} < 2^{k-1} + 2^{k-1} = 2 \cdot 2^{k-1} = 2^k = 2^{(k+2)-2}$$

and

$$f_{k+2} = f_k + f_{k+1} < 2^{(k+2)-2}. \qquad \square$$

One can in fact do better than this:

11 Theorem (Binet's Formula). *For* $n \geq 0$,

$$f_n = \frac{1}{\sqrt{5}} \left(\frac{1+\sqrt{5}}{2} \right)^n - \frac{1}{\sqrt{5}} \left(\frac{1-\sqrt{5}}{2} \right)^n. \tag{8}$$

I leave the proof by induction to the reader as an exercise.

How one arrived at this representation is not important here—and there are methods of deriving it.[8] What is important is that either theorem shows the Fibonacci sequence to grow exponentially and thus the execution of the program FIB applied to any input n to take many steps, an exponential number of calls to FIB 0 alone. The computation is *very* inefficient.

There are much more efficient procedures to calculate the Fibonacci numbers with. One is to use (8). On the *TI-83 Plus*, for example, one can go into the equation editor and enter

$$\text{Y1} = (1/\sqrt{(5)})((1+\sqrt{(5)})/2)^\wedge \text{X} - (1/\sqrt{(5)})((1-\sqrt{(5)})/2)^\wedge \text{X}$$

So long as one doesn't enter too large a number for X, this will calculate f_n. For example, it took only about a second to calculate $f_{45} = 1134903170$. Another approach, more accurate on the computer when one's language will give exact values when dealing with whole numbers and only approximations when handling $\sqrt{5}$, is to call up an auxiliary program which produces, not f_n, but either the sequence $f_0, f_1, \ldots, f_n$ or the pair f_{n-1}, f_n on input n. The first approach has the LOGO programs:

```
TO FIB :N
OUTPUT LAST FIBAUX :N
END

TO FIBAUX :N
IF :N = 0 [OUTPUT (LIST 0)]
IF :N = 1 [OUTPUT (LIST 0 1)]
LOCAL "L
MAKE "L FIBAUX (:N - 1)
OUTPUT LPUT (LAST :L + LAST BUTLAST :L) :L
END
```

For the second alternative, we might first define $f_{-1} = 1$ so that $f_1 = f_0 + f_{-1}$ holds. We then define

---

[8] We offer one method in the exercises at the end of this chapter. Another, taken from Euler, is given in the next chapter. The formula is called *Binet's Formula* after Jacques Philippe Marie Binet (1786 - 1856) who published it in 1843. It is, however, older, having been known by Euler and Daniel Bernoulli (1700 - 1782) a century earlier. Indeed, Manfred Schroeder reports on p. 65 of his *Number Theory in Science and Communication* (Springer-Verlag, Berlin, 1984) that it was discovered in 1718 by Abraham de Moivre (1667 - 1754) and proven a decade later by Nikolaus I Bernoulli (1687 - 1759).

```
TO FIB :N
IF :N = 0 [OUTPUT 0]
LOCAL "L
MAKE "L FIBAUX :N
OUTPUT (FIRST :L) + (LAST :L)
END

TO FIBAUX :N
IF :N = 0 [OUTPUT (LIST 1 0)]
IF :N = 1 [OUTPUT (LIST 0 1)]
LOCAL "L
MAKE "L FIBAUX (:N - 1)
OUTPUT (LIST (LAST :L) (FIRST :L) + (LAST :L))
END
```

One could write similar programs along these lines for the *TI-83 Plus*. On this calculator, however, there is another alternative. The sequential equation editor is designed to allow recurrent sequences u to be defined with recursive reference to u($n-1$) and u($n-2$). One sets the Mode of the calculator to Seq, then enters the equation editor and sets

nMin equal to 0

u(n) equal to u($n-1$)+u($n-2$)

u(nMin) equal to {1,0}.

Note that the list $\{1,0\}$ gives the initial values in *reverse* order. One can then calculate f_{45} by entering u(45). The calculation takes noticeably longer than does that of $Y_1(45)$. On the other hand the use of the sequential equation editor does not require one to have an explicit expression like (8) for the n-th Fibonacci number.

Speaking of (8), I would hope it hasn't escaped the reader's notice that the bases of the exponential terms are the golden ratio and its *conjugate*:

$$\phi = \frac{1+\sqrt{5}}{2}, \quad \overline{\phi} = \frac{1-\sqrt{5}}{2}.$$

This is no accident. ϕ is intimately connected with the Fibonacci numbers and I should describe two of the ways they are connected.

Consider first the continued fraction expansion of ϕ. Recall that ϕ satisfies $\phi^2 - \phi - 1 = 0$. Thus

$$\phi^2 = \phi + 1$$

$$\phi = 1 + \frac{1}{\phi}$$

$$= 1 + \cfrac{1}{1 + \cfrac{1}{\phi}}$$

$$= 1 + \cfrac{1}{1 + \cfrac{1}{1 + \cfrac{1}{\phi}}}$$

$$\vdots$$

$$= 1 + \cfrac{1}{1 + \cfrac{1}{1 + \cfrac{1}{1 + \cfrac{1}{1 + \ldots}}}}$$

The partial convergents, i.e., the simplified truncations are

$$\phi_0 = 1 = \frac{1}{1}$$

$$\phi_1 = 1 + \frac{1}{\phi_0} = 1 + \frac{1}{1} = \frac{2}{1}$$

$$\phi_2 = 1 + \frac{1}{\phi_1} = 1 + \frac{1}{2} = \frac{3}{2}$$

$$\phi_3 = 1 + \frac{1}{\phi_2} = 1 + \frac{1}{3/2} = \frac{5}{3}$$

$$\phi_4 = 1 + \frac{1}{\phi_3} = 1 + \frac{1}{5/3} = \frac{8}{5}$$

and in general we see

$$\phi_n = \frac{f_{n+2}}{f_{n+1}}. \tag{9}$$

If we formally define the sequence $\phi_0, \phi_1, \ldots$ recursively by

$$\phi_0 = 1, \quad \phi_{n+1} = 1 + \frac{1}{\phi_n},$$

the identity (9) is quickly proven by induction. We have already proven the basis. The induction step goes

$$\phi_{n+1} = 1 + \frac{1}{\phi_n} = 1 + \frac{1}{f_{n+2}/f_{n+1}}$$

$$= \frac{f_{n+2}}{f_{n+2}} + \frac{f_{n+1}}{f_{n+2}} = \frac{f_{n+2} + f_{n+1}}{f_{n+2}}$$

$$= \frac{f_{n+3}}{f_{n+2}}.$$

Another way to establish the connexion between the golden ratio and the Fibonacci sequence is to carry out the steps of the application of the Euclidean

Algorithm as applied to ϕ and 1. We referred to this in proving ϕ irrational in our discussion of the Euclidean Algorithm, but we didn't actually perform the calculations. Now we shall.

The first few steps in the application of the Euclidean Algorithm proceed as follows:

$$\phi = 1 + r_1, \text{ where } r_1 = \phi - 1$$
$$1 = r_1 + r_2, \text{ where } r_2 = 1 - r_1 = 1 - (\phi - 1) = 2 - \phi$$
$$r_1 = r_2 + r_3, \text{ where } r_3 = r_1 - r_2 = (\phi - 1) - (2 - \phi) = 2\phi - 3$$

If we write $r_0 = 1$, we find the process generates a sequence,

$$r_0 = 1$$
$$r_1 = -1 + \phi$$
$$r_2 = 2 - \phi$$
$$r_3 = -3 + 2\phi$$
$$\vdots$$

Moreover, the sequence satisfies the recurrence relation

$$r_{n+2} = r_n - r_{n+1}.$$

Knowing this, it is an easy matter to generate as many elements of the sequence we please without having to think about the Euclidean Algorithm. The process can be automated and this can be done cleverly or unthinkingly.

The unthinking[9] approach would be to go to one's calculator, store the value $(1 + \sqrt{5})/2$ in the variable F (the Latin equivalent of ϕ), go into the sequential editor and enter the values

    0 for nMin

    u($n-2$)−u($n-1$) for u(n)

    {−1+F,1} for u(nMin)

and then make a table using the **TABLE** button. As one scrolls down, one sees a nice succession of ever smaller positive decimals that don't really reveal much other than that each new 0 following the decimal point takes a long time to appear.

The clever approach is to work with $r_n = a_n + b_n\phi$ with a_n, b_n rational numbers (in fact, integers) and notice that the sequences a_n, b_n can be obtained by recurrences

$$\left. \begin{array}{ll} a_0 = 1 & b_0 = 0 \\ a_1 = -1 & b_0 = 1 \\ a_{n+2} = a_n - a_{n+1}; & b_{n+2} = b_n - b_{n+1}. \end{array} \right\} \quad (10)$$

---

[9] "Thoughtless" sounds too negative.

One can now automate the generation of the numbers a_n, b_n. One goes into the equation editor, sets 0 for $n\text{Min}$, $\mathsf{u}(n-2)-\mathsf{u}(n-1)$ for $\mathsf{u}(n)$, $\mathsf{v}(n-2)-\mathsf{v}(n-1)$ for $\mathsf{v}(n)$, $\{-1,1\}$ for $\mathsf{u}(n\text{Min})$, and $\{1,0\}$ for $\mathsf{v}(n\text{Min})$ (recalling that, for some reason or other, the initial values are reversed on the *TI-83 Plus*). One can then push the TABLE button and read off a list of values of a_n, b_n. I collect them in TABLE 5.

n	0	1	2	3	4	5	6	7	8
a_n	1	−1	2	−3	5	−8	13	−21	34
b_n	0	1	−1	2	−3	5	−8	13	−21

TABLE 5.

The values should be obvious:

$$a_n = (-1)^n f_{n+1}, \qquad b_n = (-1)^{n+1} f_n,$$

i.e.,

$$r_n = (-1)^n f_{n+1} + (-1)^{n+1} f_n \phi$$
$$= (-1)^n (f_{n+1} - f_n \phi). \tag{11}$$

12 Exercise. Prove formula (11) by induction.

Noting from our unthinking approach that r_n eventually approaches 0, we see that for large n, $f_{n+1} - f_n \phi \approx 0$, whence division by f_n yields

$$\frac{f_{n+1}}{f_n} - \phi \approx 0,$$

(where $\approx$ is the relation of approximate equality) in agreement with (9) that the ratios f_{n+1}/f_n are convergents to ϕ. In fact, because the sequence $r_0, r_1, r_2, \ldots$ is decreasing, i.e., $r_0 > r_1 > r_2 > \ldots$, and $r_0 = 1$, we have

$$1 = r_0 > |f_{n+1} - f_n \phi|,$$

whence

$$\frac{1}{f_n} > \left| \frac{f_{n+1}}{f_n} - \phi \right|,$$

and f_{n+1}/f_n is indeed a very close approximation to ϕ. As a "practical" application of this, I call the reader's attention to FIGURE 1 from the chapter on the Euclidean Algorithm. I generated this on the computer by choosing a unit and then making the horizontal lines 8.9 units long and the vertical ones 5.5 units long, giving me two nice rectangles (8.9×5.5 and 5.5×3.4) visually close to the *golden rectangle*.

There is another method of handling linear recurrences on the calculator, one that uses matrices. For those not already familiar with matrices, I shall preface the description of this method with a few remarks on what matrices are all about.

A *matrix* is a rectangular array of numbers. If the rectangle has m (horizontal) rows and n (vertical) columns, it is said to be an m-by-n matrix and the pair of numbers, written $m \times n$, is called the *dimension* (or, the *dimensions*) of the matrix. If two matrices have the same dimension, they can be added, the addition being performed elementwise. Thus, for example,

$$\begin{bmatrix} 1 & 2 & 3 \\ 4 & 5 & 6 \end{bmatrix} + \begin{bmatrix} 2 & 4 & 8 \\ 1 & 3 & 5 \end{bmatrix} = \begin{bmatrix} 1+2 & 2+4 & 3+8 \\ 4+1 & 5+3 & 6+5 \end{bmatrix} = \begin{bmatrix} 3 & 6 & 11 \\ 5 & 8 & 11 \end{bmatrix}.$$

Matrices can be multiplied by real numbers, by multiplying each element by the given number, for example,

$$5 \begin{bmatrix} 1 & 2 & 3 \\ 4 & 5 & 6 \end{bmatrix} = \begin{bmatrix} 5 \cdot 1 & 5 \cdot 2 & 5 \cdot 3 \\ 5 \cdot 4 & 5 \cdot 5 & 5 \cdot 6 \end{bmatrix} = \begin{bmatrix} 5 & 10 & 15 \\ 20 & 25 & 30 \end{bmatrix}.$$

The pertinent operation here is matrix multiplication, which arose from applications and is more complicated. To multiply two matrices, the number of elements in a row of the first matrix must equal the number of elements in a column of the second. Thus, if [A] is an $m \times n$ matrix, to form the product [A][B], [B] must have dimension $n \times p$ for some positive integer p. As for the multiplication, the product will have dimension $m \times p$ and the element of the i-th row and j-th column will be obtained by multiplying elements of the i-th row of [A] by the corresponding elements of the j-th column of [B] (i.e., first by first, second by second, etc.) and adding the products up. For example,

$$\begin{bmatrix} 1 & 2 & 3 \\ 4 & 5 & 6 \end{bmatrix} \begin{bmatrix} 3 & 6 & 9 \\ 2 & 5 & 8 \\ 1 & 4 & 7 \end{bmatrix}$$

$$= \begin{bmatrix} 1 \cdot 3 + 2 \cdot 2 + 3 \cdot 1 & 1 \cdot 6 + 2 \cdot 5 + 3 \cdot 4 & 1 \cdot 9 + 2 \cdot 8 + 3 \cdot 7 \\ 4 \cdot 3 + 5 \cdot 2 + 6 \cdot 1 & 4 \cdot 6 + 5 \cdot 5 + 6 \cdot 4 & 4 \cdot 9 + 5 \cdot 8 + 6 \cdot 7 \end{bmatrix}$$

$$= \begin{bmatrix} 3+4+3 & 6+10+12 & 9+16+21 \\ 12+10+6 & 24+25+24 & 36+40+42 \end{bmatrix}$$

$$= \begin{bmatrix} 10 & 28 & 46 \\ 28 & 73 & 118 \end{bmatrix}.$$

Note that if one reverses the order of the two matrices being multiplied in this example, the dimensions do not match and the product cannot be formed. In general, even when two matrices can be multiplied in either order, the products will not be the same: Matrix multiplication is not commutative. It is, however, associative; addition is associative and commutative; for each dimension there are an additive identity (the matrix all of whose entries are 0) and additive inverses; and the distributive laws hold.

The *TI-83 Plus* has names [A], [B], ..., [J] for 10 matrices. The matrices themselves do not exist until one creates them. This can be done from the main window by assigning a dimension to one of the names. For example, to represent the recurrence for the Fibonacci numbers, we might choose the name [F] and, since the recursion depends on 2 initial values and refers back to the 2 preceding elements of the sequence, we will need a 2×2 matrix. Thus, we enter into the calculator the following:

$$\{2,2\} \to \mathsf{dim}([F])$$

to create a 2×2 matrix. Now it doesn't matter if we use the dim(function from the MATH submenu of the MATRX menu or the corresponding function from the OPS submenu of the LIST menu— they are the same function and it can also be accessed from the CATALOG. But it is a good idea to use the MATRX button for matrix operations other than the basic arithmetic ones $(+, -, *, ^\wedge)$ for which the usual buttons work. That button opens up the NAMES submenu and one can get the name [F] from it. Do *not* try spelling it out using the [, F, and] keys.

If a matrix named [F] already existed, any elements other than those in the upper left 2×2 square are lost, and if [F] was $1 \times n$ or $m \times 1$, the new elements are all 0's. Assuming [F] previously had no value, it is now the matrix

$$\begin{bmatrix} 0 & 0 \\ 0 & 0 \end{bmatrix},$$

and is represented onscreen as

$$[\,[0\ 0]$$
$$[0\ 0]\,].$$

The matrix we want will be

$$\begin{bmatrix} 0 & 1 \\ 1 & 1 \end{bmatrix}.$$

So we have to place 1's in the correct spots. We can do this by entering

$$1 \to [F](1,2)$$
$$1 \to [F](2,1)$$
$$1 \to [F](2,2)$$

[If [X] is a matrix name, then $[X](i,j)$ is the calculator's way of referring to the element of the i-th row and j-th column of [X].] If one now enters [F], the calculator will display

$$[\,[0\ 1]$$
$$[1\ 1]\,],$$

i.e., the matrix we want.

An alternative is to use the [,] keys and enter

$$[[0,1][1,1]] \to [F]$$

directly from the home screen. Note that, while one separates the elements of the rows by commas, the rows themselves are not so separated. Also I warn again against trying to spell [F] out using the [, F,] keys.

A third alternative to creating and editing the matrix from the home screen is to go to the matrix editor by choosing the EDIT submenu from the MATRX menu and then choosing [F] as the matrix one wants to create/edit. One then keys in the dimensions and matrix elements one after another, following each entry with a press of the ENTER button, or one of the up- or down-arrow buttons. The left- and right-arrow buttons move from one digit to another in a given entry and do not enter a value. And one must be especially sure to ENTER the last entry before exiting the editor with the QUIT button.

The reason we want [F] as described is this:

$$\begin{bmatrix} 0 & 1 \\ 1 & 1 \end{bmatrix} \begin{bmatrix} x \\ y \end{bmatrix} = \begin{bmatrix} 0x + 1y \\ 1x + 1y \end{bmatrix} = \begin{bmatrix} y \\ x + y \end{bmatrix}.$$

In particular,

$$\begin{bmatrix} 0 & 1 \\ 1 & 1 \end{bmatrix} \begin{bmatrix} f_n \\ f_{n+1} \end{bmatrix} = \begin{bmatrix} f_{n+1} \\ f_n + f_{n+1} \end{bmatrix} = \begin{bmatrix} f_{n+1} \\ f_{n+2} \end{bmatrix}.$$

Thus, multiplication by [F] performs the recursive step in the recurrence. Starting with the initial value,

$$[A] = \begin{bmatrix} f_0 \\ f_1 \end{bmatrix} = \begin{bmatrix} 0 \\ 1 \end{bmatrix},$$

one can now generate successive pairs of Fibonacci numbers by performing matrix multiplications. Try entering

[F][A], [F]([F][A]) or [F]^2[A], [F]^3[A], [F]^4[A], etc.

In general

$$[F]^n[A] = \begin{bmatrix} f_n \\ f_{n+1} \end{bmatrix},$$

and in theory one could get f_n by entering

([F]^n*[A])(1,1).

For some reason I get an error message, but following ([F]^n*[A]) with Ans(1,1) works. If one plugs 45 in for n in this, one very quickly gets

$$[\, [\, 1134903170\,]$$
$$[\, 1836311903\,]\,].$$

It is noticeably faster than calculating u(45), an advantage somewhat offset by the quirk just mentioned. In writing a program, however, one could replace the desired ([F]^n*[A])(1,1) by

([F]^n*[A]) → [B]

and then use [B](1,1) whenever one wanted f_n.

Actually, we can do without [A] in this case, because $[F]^n$ is

$$\begin{bmatrix} f_{n-1} & f_n \\ f_n & f_{n+1} \end{bmatrix},$$

as one may prove by induction.

It might be amusing at this point to digress to note that this allows another derivation of the Fibonacci Addition Formula (Theorem 6). For, from

$$[F]^{m+n} = [F]^m [F]^n,$$

we can conclude

$$\begin{bmatrix} f_{m+n-1} & f_{m+n} \\ f_{m+n} & f_{m+n+1} \end{bmatrix} = \begin{bmatrix} f_{m-1} & f_m \\ f_m & f_{m+1} \end{bmatrix} \begin{bmatrix} f_{n-1} & f_n \\ f_n & f_{n+1} \end{bmatrix}$$

$$= \begin{bmatrix} f_{m-1}f_{n-1} + f_m f_n & f_{m-1}f_n + f_m f_{n+1} \\ f_m f_{n-1} + f_{m+1}f_n & f_m f_n + f_{m+1}f_{n+1} \end{bmatrix},$$

and comparing the elements of the first row, second column yields

$$f_{m+n} = f_{m-1}f_n + f_m f_{n+1}.$$

We also obtain Corollary 7 by setting $m = n$ and comparing the elements of the second row, second column:

$$f_{2n+1} = f_n^2 + f_{n+1}^2.$$

The following exercises extend our digression by applying a little more matrix theory to the Fibonacci numbers.

13 Exercise. Define the *determinant* det([X]) for

$$[X] = \begin{bmatrix} a & b \\ c & d \end{bmatrix} \text{ to be } ad - bc.$$

What is det([F])? det($[F]^n$)? Express the latter in terms of f_{n-1}, f_n, f_{n+1}. What identity follows?

14 Exercise. Evaluate $[F]^2 - [F] - I$, where I is the matrix,

$$I = \begin{bmatrix} 1 & 0 \\ 0 & 1 \end{bmatrix}.$$

How does this remind you of ϕ?

15 Exercise. The *eigenvalues* of a 2×2 matrix [X] are numbers λ such that, for some nonzero 2×1 matrix [Y],

$$[X][Y] = \lambda[Y].$$

Show that ϕ and its conjugate $\overline{\phi}$ are the eigenvalues of [F]. [Hint: Consider the preceding exercise.]

16 Exercise. Let $[X]$ be a 2×2 matrix and λ be an eigenvalue. A 2×1 matrix $[Y]$ is an *eigenvector* of $[X]$, λ if $[X][Y] = \lambda[Y]$.
i.   Show: The eigenvectors of the eigenvalues ϕ and $\overline{\phi}$ of $[F]$ are the real number multiples of the matrices

$$\begin{bmatrix} 1 \\ \phi \end{bmatrix} \text{ and } \begin{bmatrix} 1 \\ \overline{\phi} \end{bmatrix}, \text{ respectively.}$$

[Hint: Note that

$$\frac{1+\phi}{\phi} = \phi, \quad \frac{1+\overline{\phi}}{\overline{\phi}} = \overline{\phi}. \qquad]$$

ii. Show:

$$\begin{bmatrix} 0 \\ 1 \end{bmatrix} = \frac{1}{\sqrt{5}} \left(\begin{bmatrix} 1 \\ \phi \end{bmatrix} - \begin{bmatrix} 1 \\ \overline{\phi} \end{bmatrix} \right).$$

iii. Show:

$$f_n = \frac{1}{\sqrt{5}} \phi^n - \frac{1}{\sqrt{5}} \overline{\phi}^n.$$

Getting back on track, recall the sequences a_n, b_n defined earlier in (10). They share a common recurrence and thus a common matrix $[A]$:

$$\begin{bmatrix} 0 & 1 \\ 1 & -1 \end{bmatrix}.$$

And one can verify

$$\begin{bmatrix} 0 & 1 \\ 1 & -1 \end{bmatrix}^n \begin{bmatrix} 1 \\ -1 \end{bmatrix} = \begin{bmatrix} a_n \\ a_{n+1} \end{bmatrix}$$

$$\begin{bmatrix} 0 & 1 \\ 1 & -1 \end{bmatrix}^n \begin{bmatrix} 0 \\ 1 \end{bmatrix} = \begin{bmatrix} b_n \\ b_{n+1} \end{bmatrix}$$

$$\begin{bmatrix} 0 & 1 \\ 1 & -1 \end{bmatrix}^n \begin{bmatrix} 1 & 0 \\ -1 & 1 \end{bmatrix} = \begin{bmatrix} a_n & b_n \\ a_{n+1} & b_{n+1} \end{bmatrix}$$

$$\begin{bmatrix} 0 & 1 \\ 1 & -1 \end{bmatrix}^n = \begin{bmatrix} b_{n-1} & a_{n-1} \\ b_n & a_n \end{bmatrix}.$$

We can similarly represent any linear recurrence by a matrix multiplication. Consider, for example, a third order recurrence with three initial values x_0, x_1, x_2 and the recurrence relation,

$$x_{n+3} = ax_n + bx_{n+1} + cx_{n+2}.$$

A quick computation shows

$$\begin{bmatrix} 0 & 1 & 0 \\ 0 & 0 & 1 \\ a & b & c \end{bmatrix} \begin{bmatrix} x \\ y \\ z \end{bmatrix} = \begin{bmatrix} y \\ z \\ ax + by + cz \end{bmatrix},$$

and, in particular,

$$\begin{bmatrix} 0 & 1 & 0 \\ 0 & 0 & 1 \\ a & b & c \end{bmatrix} \begin{bmatrix} x_n \\ x_{n+1} \\ x_{n+2} \end{bmatrix} = \begin{bmatrix} x_{n+1} \\ x_{n+2} \\ x_{n+3} \end{bmatrix}.$$

This is handy to know as the *TI-83 Plus* is not set up to handle third order linear recurrences, but only first and second order such ones. Thanks to the built-in matrix operations, however, handling them is a snap.

7

Infinite Processes

Given numbers a, r a sequence

$$a, ar, ar^2, \ldots, ar^n$$

is called a *geometric progression*. The ancient Egyptians knew the sum

$$a + ar + ar^2 + \ldots + ar^n = \frac{ar^{n+1} - 1}{r - 1} \tag{1}$$

and Archimedes knew the infinite sum

$$a + ar + ar^2 + \ldots = \frac{-a}{r - 1} = \frac{a}{1 - r} \tag{2}$$

for $0 < r < 1$. The finite sum is easy to derive:

$$S = a + ar + ar^2 + \ldots + ar^n,$$

so

$$rS = ar + ar^2 + ar^3 + \ldots + ar^{n+1},$$

whence

$$
\begin{aligned}
rS - S &= (ar - a) + (ar^2 - ar) + \ldots + (ar^{n+1} - ar^n) \\
&= (-a + ar) + (-ar + ar^2) + \ldots + (-ar^n + ar^{n+1}) \\
&= -a + (ar - ar) + (ar^2 - ar^2) + \ldots + (ar^n - ar^n) + ar^{n+1} \\
&= -a + ar^{n+1},
\end{aligned}
$$

and

$$S = \frac{ar^{n+1} - a}{r - 1}.$$

Formally, the infinite series is obtained the same way:

$$S = a + ar + ar^2 + \ldots$$

$$rS = ar + ar^2 + ar^3 + \ldots$$
$$rS - S = -a + (ar - ar) + (ar^2 - ar^2) + \ldots$$
$$= -a,$$

and

$$S = \frac{-a}{r - 1}.$$

I stress the word "formally" here because the sum S has not been explained. For $0 < r < 1$, one can add a bunch of terms and notice, if one is representing the number in decimal form, its decimals are settling down. For example,

$$\frac{2}{5} + \left(\frac{2}{5}\right)^2 + \left(\frac{2}{5}\right)^3 + \ldots$$

has successive partial sums

.4
.56
.624
.6496
.65984
.663936
.6655744
.66622976
.666491904
.6665967616
.66663870464
.666655481856
.6666621927424

The convergence is not all that rapid, but one does see the sequence of numbers looking more and more like .6666..., which agrees with (2) for $a = .4 = r$,

$$\text{sum} = \frac{.4}{1 - .4} = \frac{.4}{.6} = \frac{4}{6} = \frac{2}{3}.$$

The first paradoxical progression one may come across would be given by $a = .9, r = .1$, where the sequence of partial sums is .9, .99, .999, ... converging to .999..., while (2) yields

$$\text{sum} = \frac{.9}{1 - .1} = \frac{.9}{.9} = 1.$$

This paradox may be explained away by performing a simple long division:

$$
\begin{array}{r}
.1\,1\,1\;\ldots \\
9\,)\,\overline{1\,.\,0\,0\,0\,0\,0} \\
\underline{9} \\
1\,0 \\
\underline{9} \\
1\,0 \\
\underline{9} \\
\ldots
\end{array}
$$

Thus $1/9 = .111\ldots$, whence $1 = 9(1/9) = 9(.111\ldots) = .999\ldots$
One may also obtain the general sum (2) by long division:

$$
\begin{array}{r}
a + ar + ar^2 + \;\ldots \\
1 - r\,)\,\overline{a} \\
\underline{a - ar} \\
ar \\
\underline{ar - ar^2} \\
ar^2 \\
ar^2 - ar^3 \\
\ldots
\end{array}
$$

This is the method used to convert rationals into decimals, which can also be infinite series. Because the decimal expansion of a rational number is periodic, it is in fact a geometric progression. Consider, for example,

$$
\frac{7}{11} = .636363\ldots
$$

$$
= \frac{63}{100} + \frac{63}{10000} + \frac{63}{1000000} + \ldots
$$

$$
= 63 \cdot \frac{1}{100} + 63 \cdot \frac{1}{100^2} + 63 \cdot \frac{1}{100^3} + \ldots
$$

We obtain the sequence .636363... from 7/11 by long division, and reverse the process by appealing to (2) to get

$$
\text{sum} = \frac{63/100}{1 - 1/100} = \frac{63/100}{99/100} = \frac{63}{99} = \frac{7}{11}.
$$

In the modern theory of limits, i.e., in the Calculus, (2) is only accepted as meaningful when $-1 < r < 1$. For, then the sum of the first n terms of the progression differs from $a/(1-r)$ by $ar^{n+1}/(r-1)$ which is very small and gets smaller as n gets larger. When $r < -1$ or $r > 1$, however, the size of this error gets larger with each increasing value of n. In the theory of limits, the infinite sum exists only if there is some such limiting number to which the partial sums tend. If the sums oscillate or grow without bound, the progression has no sum. This necessity was not always realised.

In 1696, Jakob Bernoulli discovered another paradox, one which drew a great deal of discussion down through the ages. This is the geometric progression given by $a = 1, r = -1$:

$$1 - 1 + 1 - 1 + \ldots = \frac{1}{1 - (-1)} = \frac{1}{1 + 1} = \frac{1}{2}. \tag{3}$$

The value $1/2$ was defended in various ways, including appeal to summing the geometric progression, long division of 1 by $1 + x$ (with 1 then substituted for x), and long division of 1 by $1 - x$ (with -1 then substituted for x). One also noticed that by grouping one got two different answers:

$$1 - 1 + 1 - 1 + \ldots = (1 - 1) + (1 - 1) + \ldots$$
$$= 0 + 0 + \ldots = 0$$
$$1 - 1 + 1 - 1 + \ldots = 1 + (-1 + 1) + (-1 + 1) + \ldots$$
$$= 1 + 0 + 0 + \ldots = 1.$$

Although Bernoulli first discovered the series, it is usually called the *Grandi series* after Father Guido Grandi (1671 - 1742), who mentioned it in a book a bit later and discussed it in correspondence with Gottfried Wilhelm Leibniz. Grandi remarked on how the identity

$$0 + 0 + 0 + \ldots = 1 - 1 + 1 - 1 + 1 - 1 + \ldots = \frac{1}{2}$$

reminded him of God's creation of the universe out of nothing through infinite force. Contrary to some reports he does not appear to have said it proved that God did so, but merely that it served as a reminder. It also, by the way, addresses the numerical objection to the creation that something cannot be created out of nothing.

Leibniz himself accepted the value $1/2$ for the Grandi series, justifying it by noting that the partial sums were

$$1, 1 - 1 = 0, 1 - 1 + 1 = 1, 1 - 1 + 1 - 1 = 0, \ldots$$

i.e., $1, 0, 1, 0, \ldots$ Thus half the time they were 1 and half the time 0, which means the average value was $1/2$. In modern Calculus, one points to the oscillation as proof that the series $1 - 1 + 1 - 1 + \ldots$ does not converge and therefore has no value at all. In advanced mathematical analysis, one introduces *generalised sums* and (3) actually becomes valid.

Some infinite series had in fact been introduced in India centuries earlier, and lots of them other than geometric progressions were introduced in Europe in the 17th century and thereafter. Euler presents a few in his *Elements of Algebra*[1] obtained by long division:

$$\frac{1}{1 + a} = 1 - a + a^2 - a^3 + \ldots$$

---

[1] Leonhard Euler, *Elements of Algebra*, Springer-Verlag, New York, nd, pp. 92 - 96.

$$\frac{1}{1-a} = 1 + a + a^2 + a^3 + \ldots$$

$$\frac{1}{1-a+a^2} = 1 + a - a^3 - a^4 + a^6 + a^7 - \ldots$$

Not one to reject series with no limits, he noted for this last one that, if $a = 1$ one gets

$$1 = \frac{1}{1-1+1} = 1 + 1 - 1 - 1 + 1 + 1 - \ldots$$

and that this should not be too surprising since it essentially contains two copies of Grandi's series and thus should equal

$$(1 - 1 + 1 - 1 + \ldots) + (1 - 1 + 1 - 1 + \ldots) = \frac{1}{2} + \frac{1}{2}.$$

One wonders if he noticed that interchanging the 2nd and 3rd, 6th and 7th, 10th and 11th, ... elements of the sequence results in

$$1 - 1 + 1 - 1 + \ldots = \frac{1}{2}.$$

Euler ended the discussion of such series in the *Elements of Algebra* with a few exercises expanding simple rational functions into infinite series by long division, for example:

1. Resolve $\frac{ax}{a-x}$ into an infinite series.
   Ans.

$$x + \frac{x^2}{a} + \frac{x^3}{a^2} + \frac{x^4}{a^3}, \&c.$$

4. Resolve $\frac{1+x}{1-x}$ into an infinite series.

   Ans. $1 + 2x + 2x^2 + 2x^3 + 2x^4, \&c.$

Euler had already given more sophisticated examples in his *Introduction to Analysis of the Infinite, Book I*[2]. He begins his discussion with the words

> 59. Since both rational and irrational functions of z are not of the form of polynomials $A + Bz + Cz^2 + Dz^3 + \ldots$, where the number of terms is finite, we are accustomed to seek expressions of this type with an infinite number of terms which give the value of the rational or irrational function.[3]

He starts by observing any fraction

$$\frac{a}{\alpha + \beta z}$$

is just the sum of an infinite geometric progression

---

[2] Springer-Verlag, New York, 1988. This is a recent English translation by John D. Blanton of *Introductio in analysin infinitorum* (1748).
[3] *Ibid.*, p. 50.

$$\frac{a}{\alpha} - \frac{a\beta z}{\alpha^2} + \frac{a\beta^2 z^2}{\alpha^3} - \frac{a\beta^3 z^3}{\alpha^4} + \frac{a\beta^4 z^4}{\alpha^5} - \ldots$$

One can verify this by choosing a/α for a and $-(\beta z)/\alpha$ for r in (2).
He also notes[4] that the series can be found by setting

$$\frac{a}{\alpha + \beta z} = A + Bz + Cz^2 + Dz^3 + \ldots,$$

multiplying by $\alpha + \beta z$,

$$\begin{aligned}
a &= (\alpha + \beta z)(A + Bz + Cz^2 + Dz^3 + \ldots) \\
&= \alpha A + (\alpha B + \beta A)z + (\alpha C + \beta B)z^2 + (\alpha D + \beta C)z^3 + \ldots,
\end{aligned}$$

and comparing coefficients:

$$a = \alpha A, \quad 0 = \alpha B + \beta A, \quad 0 = \alpha C + \beta B, \quad 0 = \alpha D + \beta C, \quad \ldots$$

If we follow the modern practice of writing $a_0, a_1, a_2, \ldots$ for $A, B, C, \ldots$, it is
evident that the sequence $a_0, a_1, a_2, \ldots$ is a recurrent sequence:

$$a = \alpha a_0, \quad \text{i.e., } a_0 = \frac{a}{\alpha}$$

$$0 = \alpha a_{n+1} + \beta a_n, \quad \text{i.e., } a_{n+1} = -\frac{\beta}{\alpha} a_n.$$

He refers to this method as a "continued division procedure" and shows that

$$\frac{a + bz}{\alpha + \beta z + \gamma z^2}$$

can be converted into an infinite series by such a "continued division procedure".[5] In modern notation, we would say he equates the coefficients of

$$a + bz = (\alpha + \beta z + \gamma z^2)(c_0 + c_1 z + c_2 z^2 + \ldots)$$

finds the values

$$c_0 = \frac{a}{\alpha}, \quad c_1 = \frac{b}{\alpha} - \frac{a\beta}{\alpha^2}$$

and the recurrence relation

$$\alpha c_{n+2} + \beta c_{n+1} + \gamma c_n = 0, \quad \text{i.e., } c_{n+2} = \frac{-\beta c_{n+1} - \gamma c_n}{\alpha}.$$

Euler then cites[6] the specific example

$$\frac{1 + 2z}{1 - z - z^2},$$

---

[4] *Ibid.*, p. 56.
[5] I keep the quotes so one doesn't mistake the procedure for long division.
[6] *Ibid.*, p. 53.

obtaining initial values $c_0 = 1, c_1 = 3$, and the recurrence relation $c_{n+2} = c_{n+1} + c_n$. The resulting infinite series $1 + 3z + 4z^2 + 7z^3 + 11z^4 + 18z^5 + \ldots$ shouldn't ring any bell, but the recurrence relation should sound like Pavlov's bell to us. He uses the same denominator a few pages later[7] to obtain

$$\frac{1+z}{1-z-z^2} = 1 + 2z + 3z^2 + 5z^3 + 8z^4 + \ldots$$

Simplifying further, we have

$$\frac{1}{1-z-z^2} = 1 + 1z + 2z^2 + 3z^3 + 5z^4 + \ldots$$
$$= f_1 + f_2 z + f_3 z^2 + f_4 z^3 + f_5 z^4 + \ldots, \qquad (4)$$

where $f_0, f_1, f_2, \ldots$ is the Fibonacci sequence. And multiplying by z we in fact get

$$\frac{z}{1-z-z^2} = f_0 + f_1 z + f_2 z^2 + f_3 z^3 + f_4 z^4 + \ldots$$

Another example he considers is

$$\frac{1+x}{1+x-x^2} = 1 + 0x + x^2 - x^3 + 2x^4 - 3x^5 + 5x^6 - \ldots$$
$$= f_{-1} - f_0 x + f_1 x^2 - f_2 x^3 + f_3 x^4 - f_4 x^5 + f_5 x^6 - \ldots,$$

where we define $f_{-1} = 1$.

In an earlier chapter Euler had already discussed *partial fractions*. When adding fractions, we look for a common denominator, express each fraction in terms of it, and then add the new numerators:

$$\frac{2}{3} + \frac{5}{8} = \frac{2 \cdot 8}{3 \cdot 8} + \frac{3 \cdot 5}{3 \cdot 8} = \frac{16 + 15}{24} = \frac{31}{24}.$$

In algebra we learn to do the same:

$$\frac{1}{x-2} + \frac{2}{x+3} = \frac{1}{x-2} \cdot \frac{x+3}{x+3} + \frac{2}{x+3} \cdot \frac{x-2}{x-2}$$
$$= \frac{x+3+2(x-2)}{x^2+x-6} = \frac{3x-1}{x^2+x-6}.$$

A partial fraction expansion is the reverse to this process: One starts with a fraction with a compound denominator and expands it into a sum with simpler denominators. With rational functions, if the numerator has lower degree than the denominator and the denominator has no repeated root, the process is fairly simple. For example

$$\frac{1+3x}{x^2-4} = \frac{1+3x}{(x+2)(x-2)}.$$

One assumes this to be the sum of two rational functions with denominators $x+2, x-2$, respectively, and constants for numerators and solves for the constants:

---

[7] *Ibid.*, p. 58.

$$\frac{1+3x}{(x+2)(x-2)} = \frac{A}{x+2} + \frac{B}{x-2}$$

$$1+3x = A(x-2) + B(x+2)$$
$$= (-2A+2B) + (A+B)x.$$

Comparing coefficients, we have

$$1 = -2A + 2B$$
$$3 = A + B,$$

which is easily solved:

$$A = 5/4, \quad B = 7/4.$$

Now consider the rational function

$$\frac{1}{1-z-z^2} = \frac{-1}{z^2+z-1}.$$

The quadratic formula gives us the roots to $z^2 + z - 1$ as

$$\frac{-1 \pm \sqrt{1 - 4 \cdot 1 \cdot (-1)}}{2} = \frac{-1 \pm \sqrt{5}}{2}.$$

If we recall the golden ratio ϕ and its conjugate $\overline{\phi}$,

$$\phi = \frac{1+\sqrt{5}}{2}, \quad \overline{\phi} = \frac{1-\sqrt{5}}{2},$$

we see that the roots of $z^2 + z - 1$ are $-\phi$ and $-\overline{\phi}$, yielding the factorisation

$$z^2 + z - 1 = (z - (-\phi))(z - (-\overline{\phi})) = (z+\phi)(z+\overline{\phi}).$$

Thus,

$$\frac{1}{1-z-z^2} = \frac{-1}{z^2+z-1} = \frac{-1}{(z+\phi)(z+\overline{\phi})} = \frac{A}{z+\phi} + \frac{B}{z+\overline{\phi}}$$

and we can solve for A, B:

$$-1 = A(z+\overline{\phi}) + B(z+\phi)$$
$$= A\overline{\phi} + B\phi + (A+B)z.$$

Equating coefficients, we have

$$-1 = A\overline{\phi} + B\phi, \quad 0 = A + B.$$

Thus $B = -A$ and the first equation yields

$$-1 = A\overline{\phi} - A\phi = A(\overline{\phi} - \phi)$$

$$= A\left(\frac{1-\sqrt{5}}{2} - \frac{1+\sqrt{5}}{2}\right) = A\left(\frac{-2\sqrt{5}}{2}\right),$$

and we have

$$A = \frac{1}{\sqrt{5}}, \quad B = -A = -\frac{1}{\sqrt{5}}.$$

Thus

$$\frac{1}{1-z-z^2} = \frac{1}{\sqrt{5}}\cdot\frac{1}{z+\phi} - \frac{1}{\sqrt{5}}\cdot\frac{1}{z+\overline{\phi}}$$

$$= \frac{1}{\sqrt{5}}\cdot\frac{1}{\phi+z} - \frac{1}{\sqrt{5}}\cdot\frac{1}{\overline{\phi}+z}. \tag{5}$$

The expansion of the right side into an infinite geometric progression is simpler if we write

$$\frac{1}{\phi+z} = \frac{\phi^{-1}}{1+\phi^{-1}z}, \quad \frac{1}{\overline{\phi}+z} = \frac{\overline{\phi}^{-1}}{1+\overline{\phi}^{-1}z},$$

and note that

$$\phi^{-1} = \frac{2}{1+\sqrt{5}} = \frac{2}{1+\sqrt{5}}\cdot\frac{1-\sqrt{5}}{1-\sqrt{5}} = \frac{2(1-\sqrt{5})}{1-5}$$

$$= \frac{2(1-\sqrt{5})}{-4} = -\frac{1-\sqrt{5}}{2} = -\overline{\phi},$$

and

$$\overline{\phi}^{-1} = -\phi.$$

Thus

$$\frac{1}{\phi+z} = \frac{-\overline{\phi}}{1-\overline{\phi}z} = -\overline{\phi}\left(1+\overline{\phi}z + \overline{\phi}^2z^2 + \overline{\phi}^3z^3 + \ldots\right)$$

$$= -\overline{\phi} - \overline{\phi}^2z - \overline{\phi}^3z^2 - \ldots$$

$$\frac{1}{\overline{\phi}+z} = \frac{-\phi}{1-\phi z} = -\phi\left(1+\phi z + \phi^2z^2 + \phi^3z^3 + \ldots\right)$$

$$= -\phi - \phi^2z - \phi^3z^2 - \ldots$$

and the coefficient of z^n in the series expansion of the right side of (5) is

$$\frac{1}{\sqrt{5}}\left(-\overline{\phi}^{n+1}\right) - \frac{1}{\sqrt{5}}\left(-\phi^{n+1}\right) = \frac{1}{\sqrt{5}}\phi^{n+1} - \frac{1}{\sqrt{5}}\overline{\phi}^{n+1}.$$

But we have already seen that the corresponding term of the expansion of the left side of (5) has coefficient f_{n+1}. Thus we have once again derived Binet's Formula for $n > 0$:

1 Theorem. *For all n,*

$$f_n = \frac{1}{\sqrt{5}}\left(\frac{1+\sqrt{5}}{2}\right)^n - \frac{1}{\sqrt{5}}\left(\frac{1-\sqrt{5}}{2}\right)^n. \qquad (6)$$

The case $n = 0$ can be verified by simply substituting 0 for n on both sides and noting the equality.

This calculation is rather involved and goes well beyond the middle school, but I thought it rather entertaining. It is far from rigorous, as it pays no heed to questions of convergence whence of the existence of values for the infinite series involved, and it assumes the uniqueness of the representations allowing us to equate coefficients. All of this can be justified in higher mathematics. If we wish to verify (6), we may readily do so by induction, as we did in the chapter on Fibonacci numbers, thus considering our present calculation to be an heuristic argument leading us to (6) as a conjecture to be proven.

The 17th and 18th centuries saw a lot of such heuristic arguments as mathematics saw an explosion of mathematical activity in Europe unprecedented in its history. Mathematicians summed infinite series, represented numbers as infinite products, and expanded numbers into continued fractions. In doing so they rushed forward getting new results without pausing to reflect on the correctness of their methods. They got a great deal right, but also made some errors and left a lot of gaps in their proofs to be filled in later. Leonhard Euler was a master at getting things right, though it may take quite some convincing that some of his results are not utter nonsense. Easily cited examples of these latter are his summations

$$1 + 2 + 4 + 8 + \ldots = -1$$
$$1 - 3 + 5 - 7 + \ldots = 0.$$

There are also Christian Wolff's (1679 - 1754) earlier

$$1 - 2 + 4 - 8 + \ldots = \frac{1}{3}$$
$$1 - 3 + 9 - 27 + \ldots = \frac{1}{4},$$

which Leibniz, who had accepted (3), rejected.

Here is a nice example of Euler's:

$$\frac{1}{1-x} = \frac{1}{-x+1},$$

but performing the long divisions yields

$$\frac{1}{1-x} = 1 + x + x^2 + x^3 + \ldots \qquad (7)$$

$$\frac{1}{-x+1} = -\frac{1}{x} - \frac{1}{x^2} - \frac{1}{x^3} - \ldots, \qquad (8)$$

whence

$$0 = \frac{1}{1-x} - \frac{1}{-x+1} = \ldots + \frac{1}{x^3} + \frac{1}{x^2} + \frac{1}{x} + 1 + x + x^2 + x^3 + \ldots \quad (9)$$

Is there any correct interpretation of this? The series (7) converges for all x for which $-1 < x < 1$, while that of (8) converges for all $x < -1$ or $x > 1$. There is no real value of x for which both series or the infinite sum of (9) converges.

A great deal of effort in mathematics in the 19th century went into sorting things out: deciding which methods were valid and where, and proving one's decisions to be the case. Take, for example, the continued fraction expansion of a number. For ϕ, we noted that $\phi^2 - \phi - 1 = 0$, whence

$$\phi^2 = \phi + 1$$

$$\phi = 1 + \frac{1}{\phi}$$

$$= 1 + \cfrac{1}{1 + \cfrac{1}{\phi}}$$

$$= \ldots$$

and concluded therefrom the continued fraction expansion $\phi =$

$$1 + \cfrac{1}{1 + \cfrac{1}{1 + \cfrac{1}{1 + \ldots}}} \qquad (10)$$

Now $\overline{\phi}$ also satisfies $\overline{\phi}^2 - \overline{\phi} - 1 = 0$, and

$$\overline{\phi} = 1 + \frac{1}{\overline{\phi}}$$

$$= 1 + \cfrac{1}{1 + \cfrac{1}{\overline{\phi}}}$$

$$= \ldots$$

But we cannot have both ϕ and $\overline{\phi}$ equal to the same fraction (10). Examining the convergents,

$$1, \quad 1 + \frac{1}{1} = 2, \quad 1 + \frac{1}{2} = \frac{3}{2}, \quad \ldots$$

to (10) convinces us that they do indeed tend to ϕ and not to $\overline{\phi}$. The question, however, is: why? If we start with a quadratic polynomial $x^2 + ax + b$ and derive a continued fraction expansion based on

$$x = -a - \frac{b}{x},$$

how can we be certain the convergents actually converge? And, assuming they converge to a root, which of the two roots do they converge to?

2 Exercise. Apply the procedure to the equation $x^2 - 3x + 2 = 0$. What are the roots? What value does the continued fraction expansion (i.e., its sequence of convergents) converge to? Answer the same questions for $x^2 - x - 6 = 0$.

As I said, much of the 19th century was a period of consolidation, largely a matter of clarifying vague intuitive notions like limit and convergence, rigorous definitions of which would be needed to explain which infinite sums "existed", why .999... really equalled 1, why paradoxical sums like the Grandi series could be ignored, etc.

Oversimplifying somewhat one could say this work began with an obscure Bohemian priest named Bernard Bolzano (1781 - 1848). He was a brilliant philosopher and insightful mathematician who, however, was weak on technique. He gave a fairly clear and correct definition of a continuous function— a function which preserves limits: if f is a function and $x_0, x_1, \ldots$ is a sequence of approximations to x, then $f(x_0), f(x_1), \ldots$ tends to $f(x)$. In 1817 he gave a correct[8] proof of the *Intermediate Value Theorem*: If f is continuous on an interval $[a, b]$ and $f(a) < 0 < f(b)$, then $f(c) = 0$ for some c in the interior (a, b) of $[a, b]$. Bolzano, however, lived in Prague, not then a major mathematical centre, and he was overlooked. One reason was Cauchy.

Augustin Louis Cauchy (1789 - 1857) was a prolific mathematician living in Paris, at the time one of the great centres of mathematics. In 1821 he published a book, *Cours de analyse* [*Course of analysis*], providing similar rigorous definitions of the key concepts of the Calculus. This work was polished into a more definitive final form by Karl Weierstrass (1815 - 1897).

One big problem remaining was a characterisation of or foundation for real numbers. With a continued fraction,

$$a_0 + \cfrac{1}{a_1 + \cfrac{1}{a_2 + \cfrac{1}{a_3 + \ldots}}}$$

for example, if the limit were known, Bolzano-Cauchy-Weierstrass had given the tools for proving the fraction converged to that limit. But if one did not know the limit, there were no tools to prove its existence. The problem was one of characterising the real numbers in some useful way.

A start on the problem of the real numbers was made by Bolzano in the early 1830's, but was left unfinished with his death in 1848 and went unknown until the 1960's. Weierstrass took up the problem and came up with a solution around 1840. He lectured on it in Berlin, but never published his results. In 1858 Richard Dedekind came up with a much simpler solution, one that remains popular in expositions today. Both the solutions by Weierstrass and Dedekind

---

[8] Not everyone agrees. However, with his characterisation of the real numbers, the proof is indeed correct.

went unpublished until 1872, when another solution in three variations was published by Charles Méray (1835 - 1911), Eduard Heine (1821 - 1881), and Georg Cantor (1845 - 1918). Other than Bolzano's characterisation, which was intended to be a description of the real numbers, these were presented as infinitistic constructions of the real numbers from the rationals. In each case, a real number is somehow constructed from an infinite collection of rationals.

This probably sounds a bit mysterious, but one of these constructions, that by Dedekind, is easy to describe. The rational numbers have a lot of gaps in them. $\sqrt{2}$ is not rational and it divides the rational numbers into two non-overlapping sets—those rational numbers less than $\sqrt{2}$,

$$A = \{x \mid x \text{ is rational } \& \ x < \sqrt{2}\,\},$$

and those greater than $\sqrt{2}$,

$$B = \{x \mid x \text{ is rational } \& \ x > \sqrt{2}\,\}.$$

Note that A, B are nonempty ($1 \in A, 2 \in B$), have no elements in common, exhaust all the rationals, and are such that every element of A is less than every element of B. Such a pair of sets he called a *cut*, and nowadays it is called a *Dedekind cut* in his honour. Dedekind obtained the real numbers from the rationals by adding a new real number for every cut that was not already determined by a rational number, i.e., for which there was no greatest element of A or least element of B

This was not the end of the story. A precise and rigorous description of the real numbers and what it meant for a function to be continuous made possible the construction in the latter decades of the 19th century of *monsters*—curves with no tangents or arc lengths, bounded sets with no area, etc. And the introduction of sets made possible the discovery of new paradoxes concerning the infinite. Paradoxes involving infinite sets go back to the Greeks, who solved them by denying the existence of completed infinite sets, allowing only the *potential infinite*: A set could fail to be finite because, at any given time, it could be extended by the addition of an unbounded number of new elements, but at no time did it actually have an infinite number of such. Its infiniteness was merely a potentiality, not an actualised fact.

In the 17th century actual infinites were making an appearance and Galileo noted a couple of paradoxes connected with them. And in the 19th century, Bolzano collected a number of paradoxes of the infinite together, ranging from Grandi's sequence to "Galileo's paradox", in a slim monograph posthumously published in 1851 under the title *Paradoxien des Unendlichen* [*Paradoxes of the Infinite*]. With Dedekind's cuts, for example, actually infinite sets became more or less concrete objects. Heinrich Weber (1842 - 1913), in expounding Dedekind's construction of the real numbers, took the real numbers to be the cuts themselves, thus not only accepting infinite sets as completed objects, but sets of infinite sets as such. By this time Cantor had begun developing set theory dealing extensively with infinite sets and various manifestations of infinitude. While Galileo's old paradoxes were easily cleared up, new paradoxes arose,

the most famous, and simplest, being *Russell's paradox*, named after Bertrand
Russell (1872 - 1970), who discovered it in 1902, though we now know it had
already been discovered by Ernst Zermelo (1871 - 1953) in the 1890's: Consider
the set R of all sets which are not elements of themselves:

$$R = \{x \mid x \text{ is not an element of } x\}.$$

The question is, is R an element of itself? If the answer is "yes", then by definition R does not belong to itself; and if the answer is "no", then by definition
R should belong to itself. Either possibility leads to a contradiction.

Histories of mathematics in the 20th century make much of these paradoxes,
and they indeed occupied many philosophers and some mathematicians, but
the fact is that, unlike the problems concerning limits of series and functions,
the new paradoxes concerned only the fringes of mathematics, not the mainstream. While many mathematicians of the early decades of the 20th century
occasionally considered them, few worked full time on them. The main task of
consolidation had been completed in the 19th century and 20th century mathematics was largely devoted to abstract generalisation, with occasional new
fields like mathematical logic, computer science, or game theory opening up.

8

Square Roots

When I was in elementary school we were taught how to find square roots by hand. It was a laborious procedure involving making an initial guess and then improving one's estimate one digit at a time by considering pairs of digits of the original number. Needless to say, I never remembered it. But I did look it up[1] and can report that it is based on the Binomial Theorem. The idea is very simple. Given a nonsquare number d the square root of which one wishes to find, one starts by choosing $a = [\sqrt{d}\,]$ = the greatest integer $\leq \sqrt{d}$, and then one finds the largest b such that $(a + b)^2 = a^2 + 2ab + b^2 < d$, i.e.,

$$(2a + b) \cdot b < d - a^2. \tag{1}$$

When I was a student, we did it in an imitation of long division as in FIGURE 1 on the next page. The explanation of the Figure is this. One groups the digits in pairs starting at the decimal point and working away from it. The leftmost grouping is either a pair (if the integral part of the number has an even number of digits as in our example) or a singleton (should the integral part possess an odd number of digits).[2] In our example, which I borrowed from Chabert's *History of Algorithms*, $d = 189574$ and the first pair is 18. The integral part of the square root of 18 is 4, which we place above the 18. If we subtract $4^2 = 16$ from 18 we get 2. Bring down the 95 and we have replaced 1895 with $1895 - 40^2 = 1895 - 1600 = 295$. By (1) we want the largest b such that

$$(2 \cdot 40 + b)b < 1895 - 1600 = 295.$$

From $(80 + b)b = 80b + b^2 < 295$, we conclude

---

[1] It can be found in Chapter 7, section 3, of Jean-Luc Chabert (ed.), *A History of Algorithms; From the Pebble to the Microchip*, Springer-Verlag, Berlin, 1999 (pp. 205 - 208). This is an interesting history/source book covering many algorithms that might be of interest in the middle school as well as many from more advanced mathematics.

[2] The grouping consists basically in writing d in base 100. Placing the digits of $\sqrt{d}$ above the pairs is writing $\sqrt{d}$ in base 10.

$$
\begin{array}{r}
4\ \ 3\ \ \ 5\ .\ \ 4\ \ \ 0\ \ \ \ 0\ \ \ \ 9 \\
\sqrt{\ 18\ \ 95\ \ 74\ .\ 00\ \ 00\ \ 00\ \ 00} \\
\underline{16} \\
2\ \ 95 \\
\underline{2\ \ 49} \\
4\ 6\ \ 7\ 4 \\
\underline{4\ 3\ \ 2\ 5} \\
3\ \ 4\ 9\ \ \ \ 0\ 0 \\
\underline{3\ \ 4\ 8\ \ \ \ 1\ 0} \\
8\ 4\ \ 0\ 0 \\
\underline{0} \\
8\ 4\ \ 0\ 0\ \ 0\ 0 \\
\underline{0} \\
8\ 4\ \ 00\ \ 00\ \ 0\ 0 \\
7\ 8\ \ 37\ \ 20\ \ 8\ 1 \\
\underline{} \\
5\ \ 6\ 2\ \ 79\ \ 19
\end{array}
$$

FIGURE 1. Finding a Square Root

$$
b + \frac{b^2}{80} < \frac{295}{80}.
$$

In general b is at most $\left[\frac{d-(10a)^2}{20a}\right]$. In the present case,

$$
b = \left[\frac{295}{80}\right] = [3.6875] = 3.
$$

So we place 3 above the 95 and subtract $3(80 + 3) = 249$ from 295 to get 46. Now bring down the 74, and consider $(2 \cdot 430 + b)b$: Choose

$$
b = \left[\frac{4674}{860}\right] = [5.43\ldots] = 5
$$

and place the 5 above the 74. And continue in this manner.

Finding the square root of 2 affords us a simple example where we have to subtract 1 in determining b. Consider FIGURE 2. We write 2 in base 100, look at the first base-100-digit 2 and note that the greatest integral square less than or equal to 2 is 1. Its square root is 1 and this is put above the 2, and its square is subtracted from 2 to get 1. We then place the two 0's after the result of the subtraction. $a(=1)$ now becomes $10a(=10)$ and we look at

$$
\left[\frac{200 - 10^2}{20}\right] = 5.
$$

This is *not* our choice of b because

$$
5 + \frac{5^2}{2a} = 5 + \frac{25}{20} > 5 = \frac{200 - 10^2}{20}.
$$

$$
\begin{array}{r}
1 \, . \quad 4 \quad 1 \quad 4 \quad 2 \\
\sqrt{\,\overline{2 \, . \, 00 \ 00 \ 00 \ 00}} \\
\end{array}
$$

```
      1 .   4   1   4   2
  _____
 √ 2 . 00 00 00 00
   1   00
       9 6
     _____
       4  0 0
       2  8 1
     _____
       1 19  0 0
       1 12  9 6
          _____
          6 04  0 0
          5 65  6 4
             _____
             38  36
```

FIGURE 2.

Instead we choose $b = 4$. We now subtract $2ab + b^2 = 80 + 4^2 = 96$ from 100 to get 4. The next step we are looking at

$$20000 - 140^2 = 400,$$

divided by $2 \cdot 140 = 280$:

$$\left\lfloor \frac{400}{280} \right\rfloor = [1.428\ldots] = 1,$$

and we take $b = 1$. Then

$$\left\lfloor \frac{11900}{2 \cdot 1410} \right\rfloor = \left\lfloor \frac{11900}{2820} \right\rfloor = [4.219\ldots] = 4$$

and we can take $b = 4$. And the process continues.

Variants of this method go back to 3rd century China and were also known in India, and in the Middle Ages throughout the Mediterranean.

By the mid-13th century, Qín Jiǔsháo used another method now called *Horner's method* after William George Horner (1786 - 1837), who, to oversimplify the history a bit, rediscovered the method in the 19th century. This was a general method of determining a solution to a polynomial equation one decimal at a time. Let $d = 189574$ and consider the polynomial $P(X) = X^2 - 189574$. As d has 6 digits, its square root ought to have 3, so one makes a table. Noting

X	100	200	300	400	500	600
$P(X)$	-179574	-149574	-99574	-29574	60426	$\ldots$

that $P(X)$ changes sign between $X = 400$ and 500, one sees that the root lies between 400 and 500 and can be written as $X = Y + 400$ for some number Y less than 100. Substituting $Y + 400$ for X in $P(X)$ yields

$$Q(Y) = (Y + 400)^2 - 189574 = Y^2 + 800Y + 160000 - 189574$$

$$= Y^2 + 800Y - 29574.$$

One now makes another table: And one sees a sign change between 30 and 40.

X	0	10	20	30	40	50
$Q(X)$	-29574	-21474	-13174	-4674	4026	$\ldots$

So one can set $Y = Z + 30$,

$$R(Z) = (Z + 30)^2 + 800(Z + 30) - 29574$$
$$= Z^2 + 60Z + 900 + 800Z + 24000 - 29574$$
$$= Z^2 + 860Z - 4674.$$

Again one makes a table And one sees that Z lies between 5 and 6. One can

Z	0	1	2	3	4	5	6	7
$R(Z)$	-4674	-3813	-2950	-2085	-1218	-349	522	$\ldots$

now write $Z = W + 5$ and choose $S(W) = (W + 5)^2 + 860(W + 5) - 4674$ and continue.

One might notice that the coefficients of the successive polynomials were encountered in the 3rd century algorithm and that the successive sign changes took place near the ratios 29574/800 and 4674/860, that is (ignoring the sign) the ratios of the coefficients of the constant terms to those of the linear terms. Thus, one would not have to calculate $Q(0), Q(10), Q(20), \ldots$, but could find $29574/800 = 36.9675$ and know that Y was somewhere between 30 and 40. And since $4674/860 \approx 5.43$, Z lies somewhere between 5 and 6. Doing this, the method is essentially the same as the earlier one, but written in a more easily memorable way.

Through each cycle this algorithm brings one a single additional digit in the square root: $400, 400 + 30 = 430, 430 + 5 = 435, \ldots$ One can get greater efficiency by choosing the ratios themselves. In the 17th century, Newton would have started with the approximation 400 to the square root of 189574. He would then have chosen

$$Q(Y) = (Y + 400)^2 - 189574 = Y^2 + 800Y - 29574$$

as before, and then considered $29574/800 = 36.9675$, perhaps rounded[3] this to 37 and set $Y = Z + 37$ to get

---

[3] There is a tradeoff here. Rounding makes the multiplications easier, but rounding too much loses accuracy and a greater number of steps will be involved to reach a desired degree of accuracy.

$$R(Z) = (Z + 37)^2 + 800(Z + 37) - 29574$$
$$= Z^2 + 74Z + 1369 + 800Z + 29600 - 29574$$
$$= Z^2 + 874Z + 1395.$$

One now notes $1395/874 \approx 1.59610984$ and, since the constant term is positive, one regards this as an excess and subtracts it from 437: 435.4038902. If we round this off to 435.40, we are accurate to two more decimal places than the same number of cycles of the old Chinese method. Of course, with the old Chinese method we knew how many significant digits were accurate and with Newton's method we have to make a further analysis or carry out another step or two to see where there is no change.

If one is not wed to decimal representations, there are other ways of finding rational approximations to square roots. A very simple method that may go back to the Babylonians (2000 - 1700 B.C.), who used many approximations to square roots and thus may be assumed to have employed some method, was used as the basis of an iterative procedure by Hero of Alexandria (*fl. c.* 50 A.D.). He described it as follows:

> Since 720 has not a rational square root, we shall make a close approximation to the root in this manner. Since the square nearest to 720 is 729, having a root of 27, divide 27 into 729; the result is $26\frac{2}{3}$; add 27; the result is $53\frac{2}{3}$. Take half of this; the result is $26\frac{1}{2} + \frac{1}{3}$ ($= 26\frac{5}{6}$). Therefore the square root of 720 will be very nearly $26\frac{5}{6}$. For, $26\frac{5}{6}$ multiplied by itself gives $720\frac{1}{36}$; so that the difference is $\frac{1}{36}$. If we wish to make the difference less than $\frac{1}{36}$, instead of 729 we shall take the number now found, $720\frac{1}{36}$, and by the same method we shall find an approximation differing by much less than $\frac{1}{36}$.[4]

As was the practice, Hero explains the algorithm by giving a single instance, but the method is clear. To find $\sqrt{d}$, start with a number x close to $\sqrt{d}$, i.e., a number x for which x^2 is close to d. Since x is close to $\sqrt{d}$, d/x is also close to $\sqrt{d}$ and, insofar as $x \neq \sqrt{d}$, $\sqrt{d}$ will lie between x and d/x. So take the average as a next approximation:

$$x' = \frac{1}{2}\left(x + \frac{d}{x}\right).$$

The step can be iterated. Thus, let x_0 be chosen so that x_0^2 is close to d and then successively obtain $x_1, x_2, x_3, \ldots$ by setting

$$x_{n+1} = \frac{1}{2}\left(x_n + \frac{d}{x_n}\right).$$

With our example $d = 189574$, starting at $x_0 = 400$, the succession goes:

400

---

[4] Chabert, *op. cit.*, p. 202; and Ivor Thomas, *Selections Illustrating the History of Greek Mathematics, II: From Aristarchus to Pappus* (aka: *Greek Mathematical Works*, vol. II), Harvard University Press, Cambridge (Mass.), 1961, p. 471.

$$200 + \frac{189574}{800} = 200 + 236 + \frac{774}{800} = 436 + \frac{387}{400}$$

$$218\frac{387}{800} + \frac{189574}{2\left(436 + \frac{387}{400}\right)} = 218\frac{387}{800} + \frac{189574 \cdot 200}{436 \cdot 400 + 387}$$

$$= 218 + \frac{387}{800} + \frac{37914800}{174787}$$

$$= 218 + \frac{387}{800} + 216 + \frac{160808}{174787}$$

$$= 434 + \frac{67642569 + 128646400}{139829600}$$

$$= 434 + \frac{196288969}{139829600}$$

$$= 435 + \frac{56459369}{139829600}$$

In decimals, these values are

400

436.9675

435.4037726561...

Foregoing the pleasure of calculating with fractions by hand and entering 400, ENTER, .5(Ans + 189574/Ans) in the calculator, successive hits on the ENTER button yield the sequence of improvements:

436.9675

435.4037727

435.4009646

435.4009646

and these last two agree with the calculator's value of $\sqrt{189574}$.

In his commentary on Ptolemy's *Almagest*, Theon of Alexandria (*fl.* 2nd half of 4th century A.D.) explains how Ptolemy found $67°4'55''$ as the square root of $4500°$, i.e.,

$$\sqrt{4500} \approx 67 + \frac{4}{60} + \frac{55}{60^2}$$

(This is $67.0819\overline{4}$ and its square is approximately 4499.98727.), using a variant of this method, which he explained geometrically via reference to the geometric algebra of Book II of *The Elements*.

Hero's method converges very rapidly, a fact that can be verified today by any student of the Calculus. For, it turns out to be identical with the workings of the *Newton-Raphson Method*, the convergence of which is analysed in a Calculus course. The Newton-Raphson Method is explained geometrically. To find $\sqrt{d}$, one tries to find a root of the function $f(x) = x^2 - d$, i.e., a point where the parabola $y = x^2 - d$ crosses the x-axis. To do this, one starts with an initial estimate a for $\sqrt{d}$ and improves it by drawing the tangent to the curve at $(a, f(a))$ (See FIGURE 3.) and choosing the point b where the tangent crosses

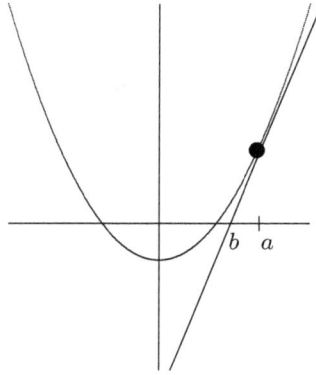

FIGURE 3. Newton-Raphson Method

the x-axis as the next approximation. This is doable because the equation of the tangent line is easy to find and is linear. From the Calculus one knows that the slope of the tangent to $y = x^2 - d$ at the point $(a, a^2 - d)$ is just $2a$. We also know that $(a, a^2 - d)$ lies on the tangent line, whence its equation is

$$\frac{y - (a^2 - d)}{x - a} = 2a,$$

i.e.,

$$y = a^2 - d + 2a(x - a).$$

This crosses the x-axis when $y = 0$:

$$0 = a^2 - d + 2a(x - a)$$
$$d - a^2 = 2ax - 2a^2$$
$$2ax = d + a^2$$
$$x = \frac{a^2 + d}{2a} = \frac{1}{2}\left(a + \frac{d}{a}\right). \tag{2}$$

1 Remark. One doesn't need the Calculus to determine the slope of a tangent line to a parabola $y = x^2 - d$ at some point $(a, a^2 - d)$. There are only two straight lines passing through the point with no other points of intersection, namely the vertical line $x = a$, which has no slope, and the tangent line, which has a finite slope. Suppose a line,

$$y = mx + b$$

passing through the point $(a, a^2 - d)$ has $(a + h, a^2 + 2ah + h^2 - d)$ as another point of intersection with the parabola. Now the point $(a, a^2 - d)$ is on both the parabola and the line, whence

$$a^2 - d = ma + b. \tag{3}$$

Similarly, $(a + h, a^2 + 2ah + h^2 - d)$ lies on both curves:

$$a^2 + 2ah + h^2 - d = m(a + h) + b. \tag{4}$$

Subtracting (3) from (4) yields

$$2ah + h^2 = mh,$$

i.e., if $h \neq 0$,

$$2a + h = m.$$

So the slope of any non-tangent line is not $2a$, whence $2a$ must be the slope of the tangent line.

Earlier we came across the use of solutions x, y to the Pell equation,

$$X^2 - dY^2 = 1,$$

to obtain the approximation x/y to $\sqrt{d}$. However, we also experienced some difficulty in finding x, y. A fairly intelligible approach, not necessarily the most efficient, is to apply the Euclidean Algorithm. Let me begin with the very simple example of $d = 2$, i.e., we apply the Euclidean Algorithm to $\sqrt{2}, 1$:

$$\sqrt{2} = 0 + 1\sqrt{2} = 1 \cdot 1 + (\sqrt{2} - 1) = 1(1 + 0\sqrt{2}) + (-1 + 1\sqrt{2})$$
$$1 = 1 + 0\sqrt{2} = 2(-1 + 1\sqrt{2}) + (3 - 2\sqrt{2})$$
$$-1 + 1\sqrt{2} = 2(3 - 2\sqrt{2}) + (-7 + 5\sqrt{2})$$
$$3 - 2\sqrt{2} = 2(-7 + 5\sqrt{2}) + (17 - 12\sqrt{2})$$
$$-7 + 5\sqrt{2} = 2(17 - 12\sqrt{2}) + (-41 + 29\sqrt{2})$$
$$17 - 12\sqrt{2} = 2(-41 + 29\sqrt{2}) + (99 - 70\sqrt{2})$$

Among the remainders, one might recognise the pairs $(3, 2), (17, 12)$, and $(99, 70)$ as solutions to $X^2 - 2Y^2 = 1$ and the ratios $3/2, 17/12$, and $99/70$ as approximations,

$$\frac{3}{2} = 1.5, \quad \frac{17}{12} = 1.41\overline{6}, \quad \frac{99}{70} \approx 1.414285714,$$

to $\sqrt{2} \approx 1.414213562$. It turns out the other pairs satisfy $X^2 - 2Y^2 = -1$ and are also good approximations to $\sqrt{2}$:

$$\frac{1}{1} = 1, \quad \frac{7}{5} = 1.4, \quad \frac{41}{29} \approx 1.413793103.$$

The successive solutions to $X^2 - 2Y^2 = 1$ are progressively better approximations from above, while the solutions to $X^2 - 2Y^2 = -1$ are progressively better approximations from below.

Moreover, the quotients $1, 2, 2, 2, \ldots$ of the divisions are just the numbers occurring in the simple continued fraction expansion of $\sqrt{2}$:

$$\sqrt{2} = 1 + \cfrac{1}{2 + \cfrac{1}{2 + \cfrac{1}{2 + \ldots}}}$$

This generalises to some extent. One can start with any d not a perfect square and find sequences $q_0, q_1, q_2, \ldots, r_0, r_1, r_2, \ldots,$

$$\sqrt{d} = q_0 \cdot 1 + r_0, \quad 0 < r_0 < 1$$
$$1 = q_1 r_0 + r_1, \quad 0 < r_1 < r_0$$
$$r_0 = q_2 r_1 + r_2, \quad 0 < r_2 < r_1$$

$$\vdots$$

where each q_n is a whole number and each r_n is of the form $a_n + b_n \sqrt{d}$ with a_n, b_n integral of opposite signs. The sequences satisfy

$$\sqrt{d} = q_0 + \cfrac{1}{q_1 + \cfrac{1}{q_2 + \cfrac{1}{q_3 + \ldots}}},$$

and
$$a_n^2 - db_n^2 = c_n, \text{ for some } c_n.$$

That $a_n^2 - db_n^2 = $ something is, of course, no revelation, but there is a pattern to the sequence $c_0, c_1, c_2, \ldots$, which eventually hits on another $c_n = 1$ after $c_0 = 1$, after which point the numbers repeat: $c_0, c_1, \ldots, c_n (= c_0), c_1, \ldots, c_n, \ldots$ Likewise, the sequence $q_0, q_1, \ldots$ is periodic. One is most readily convinced of this by examining numerous examples. Such an examination would be too lengthy a digression, so I postpone it, pausing only to note that applying the technique to our ever present $d = 189574$ yields the estimate:

$$\sqrt{189574} \approx \frac{105822773647803734751817738262712499594373305 1}{2430467138214450807862160409768112126250670} \tag{5}$$

$$\approx 435 + \frac{97453135475124609813760437799622102469160 1}{2430467138214450807862160409768112126250670},$$

the square of which differs from 189574 by a mere

$$\left(\frac{1}{2430467138214450807862160409768112126250670} \right)^2 ,$$

which I haven't multiplied out because the denominator of the square is too large to fit on a single line. I leave it to the reader to convert the estimate to decimal form.

There is yet another method of finding square roots that should be mentioned. This is Newton's Binomial Theorem. The ordinary Binomial Theorem expressing a positive integral power of a binomial in terms of the binomial coefficients,

$$(a+b)^n = \binom{n}{0}a^n + \binom{n}{1}a^{n-1}b + \ldots + \binom{n}{n}b^n$$

$$= \sum_{k=0}^{n} \binom{n}{k}a^{n-k}b^k, \tag{6}$$

where

$$\binom{n}{k} = \frac{n(n-1)\cdots(n-k+1)}{k(k-1)\cdots 1}, \tag{7}$$

was known in China centuries before Blaise Pascal rediscovered and popularised it in Europe. Noting that

$$\binom{n}{k} = 0 \text{ for } n > k,$$

Newton observed that (6) could be written as an infinite sum,

$$(a+b)^n = \sum_{k=0}^{\infty} \binom{n}{k}a^{n-k}b^k.$$

Combining this with the observation that the right-hand side of (7) can be evaluated for any rational value of n, and adding a few other considerations, Newton proclaimed his own generalisation of the Binomial Theorem:

2 Theorem (Newton's Binomial Theorem). *For any rational exponent α,*

$$(a+b)^\alpha = \sum_{k=0}^{\infty} \binom{\alpha}{k}a^{\alpha-k}b^k \tag{8}$$

$$= a^\alpha + \frac{\alpha}{1}a^{\alpha-1}b + \frac{\alpha(\alpha-1)}{2\cdot 1}a^{\alpha-2}b^2 + \ldots$$

As a particular case, for $\alpha = 1/2$ we have

$$\sqrt{a+b} = (a+b)^{1/2} = a^{1/2} + \frac{1}{2}a^{-1/2}b + \frac{\frac{1}{2}(-\frac{1}{2})}{2\cdot 1}a^{-3/2}b^2 + \ldots$$

$$= a^{1/2} + \frac{1}{2}a^{-1/2}b - \frac{1}{8}a^{-3/2}b^2 + \frac{1}{16}a^{-5/2}b^3 - \frac{5}{128}a^{-7/2}b^4 + \ldots \tag{9}$$

To guarantee one is not merely shifting the difficulty of finding $\sqrt{a+b}$ to that of finding $\sqrt{a}$, one chooses a to be a perfect square close to some given value of d. And, to guarantee that the series (9) converges, i.e., that the partial sums settle down, one must also make sure that $b = d-a$ is relatively small compared to a.

3 Example. To explain what I mean by the convergence problem, consider the cases $d = 2$ and $d = 3$. In each case, we take $a = 1$. For $d = 2$, we have $b = 1$ and

$$\sqrt{2} = \sqrt{1+1} = 1 + \frac{1}{2} - \frac{1}{8} + \frac{1}{16} - \frac{5}{16} + \frac{7}{256} - \cdots$$
$$= 1 + .5 - .125 + .0625 - .0390625 + .02734395 - \cdots$$

with the partial sums

$$1, 1.5, 1.375, 1.4375, 1.3984375, 1.42578125, \ldots,$$

which are slowly converging to $\sqrt{2} \approx 1.414213562$. For $d = 3$, we have $b = 2$ and

$$\sqrt{3} = \sqrt{1+2} = 1 + 1 - .5 + .5 - .625 + .875 - 1.3125 + 2.0675 - \cdots,$$

with the partial sums

$$1, 2, 1.5, 2, 1.375, 2.25, .9375, 3, -3.515625, 5.234375,$$
$$- 4.26171875, 12.140625, -16.5631167656, \ldots,$$

with an ever increasing oscillation. The series does not converge to anything.

It turns out that the series in question will converge so long as

$$-1 < \frac{b}{a} \leq 1,$$

the closer b/a is to 0, the swifter the convergence.

One can explore this easily on one's calculator. First, it is convenient to factor the a out:

$$\sqrt{a+b} = \sqrt{a}\sqrt{1 + \frac{b}{a}} = a^{1/2}\left(1 + \frac{b}{a}\right)^{1/2}$$
$$= a^{1/2}(1+x)^{1/2}, \text{ writing } x = \frac{b}{a}$$
$$= a^{1/2}\left(\binom{1/2}{0} + \binom{1/2}{1}x + \binom{1/2}{2}x^2 + \cdots\right) \tag{10}$$

On the *TI-83 Plus* one can store $\sqrt{a}$ in the variable A, b in B, and then store b/a in X. For our repeated example $d = 189574$, one would punch the following commands into the calculator:

```
400 →A
189574−400²→ B
B/400² →X.
```

The next step is to make sure the calculator is in sequential mode and to go into the Equation Editor and enter

0 for nMin
$-(2n-3)*X*u(n-1)/(2n)$ for $u(n)$
1 for $u(n$Min$)$.

[The reason for the second step is that

$$\binom{\alpha}{k+1} = \frac{\alpha(\alpha-1)\cdots(\alpha-(k+1)+1)}{(k+1)\cdots1}$$
$$= \frac{\alpha(\alpha-1)\cdots(\alpha-k+1)}{k\cdots1} \cdot \frac{\alpha-k}{k+1} = \binom{\alpha}{k}\frac{\alpha-k}{k+1}.$$

In particular,

$$\binom{1/2}{n} = \frac{\frac{1}{2}-(n-1)}{n}\binom{1/2}{n-1} = \frac{\frac{1-2n+2}{2}}{n}\binom{1/2}{n-1} = -\frac{2n-3}{2n}\binom{1/2}{n-1}.$$

Thus one recursively finds the individual summands of the series in (10) by:

$$\text{1st term} = \binom{1/2}{0} = 1$$
$$n\text{th term} = -\frac{2n-3}{2n} \cdot ((n-1)\text{-st term}) \cdot x. \qquad]$$

Having done this, one can now generate a few lists. For the list of terms, one picks a large number, say 15, and stores $u(0),\ldots,u(15)$ in L_1:

seq$(u(n),n,0,15)\rightarrow L_1$,

or even

u$(0,15)\rightarrow L_1$.

Then one takes the cumulative sums of L_1 and stores them in L_2:

cumSum$(L_1)\rightarrow L_2$,

and multiplies this by $\sqrt{a}$ and stores the results in L_3:

A$*L_2 \rightarrow L_3$.

One can then read the partial sums from L_3:

400, 436.9675, 435.2592549, 435.4171288, 435.3988907, 435.4012504
435.4009233, 435.4009708, 435.4009637, 435.4009648,
435.4009646, ..., 435.4009646,

the omitted values being identical to 435.4009646, the value given by the calculator when $\sqrt{(189574)}$ is entered.

4 Exercise. Use your calculator to approximate $\sqrt{2}$ using this method, writing $2 = 1.96 + .04 = (1.4)^2 + .04$.

For this Exercise, if you carried out the above computation while reading, you already have u entered in the Equation Editor and have only to store new values 1.4 in A, .04 in B, and B/A^2 in X before performing the list operations. If one wanted to experiment a bit, a program would be nice. Here is one that assumes values for $\sqrt{a}, b$ have already been stored in A and B:

```
PROGRAM:ROOTAPLB    ("PL" for "Plus")
:B/A² →X
:u(0,15)→L₁
:cumSum(L₁)→L₂
:A∗L₂ →L₃
```

This assumes 15 is a sufficient number of terms. You can replace it by a variable, say N, and store a positive integer in N before running the program. If you imagine yourself returning to the program at a later date after which you may have been exploring with other sequences u, you might prefer the following program which replaces the existing sequence u by the one just described:

```
PROGRAM:ROOTAPLB
:B/A² →X
:"−(2n−3)∗X∗u(n−1)/(2n)"→u
:0→ nMin
:1→u(nMin)
:u(0,15)→L₁
:cumSum(L₁)→L₂
:A∗L₂ →L₃
```

The hard part of entering this program into the *TI-83 Plus* is finding nMin and u(nMin) in the calculator. They are hidden away in the U/V/W submenu of the WINDOWS menu accessed from the variables menu arrived at by pressing the VARS button. And if you don't already have the calculator in sequential mode, you have to use the CATALOG button and jump to n by hitting the N button before hitting ENTER.

For a lot of exploration, the larger screen of the computer is convenient. Here is a LOGO program for the task.

```
TO ROOT.A.PLUS.B :A :B :N
LOCAL "X
MAKE "X :B/(:A ∗ :A)
OUTPUT SCALAR.MULT :A (CUM.SUM NEWTON.LIST :X)
END

TO NEWTON.LIST :X :N
IF :N = 0 [OUTPUT (LIST 1)]
LOCAL "NL
MAKE "NL (NEWTON.LIST :X (:N −1 ))
OUTPUT LPUT (−(2 ∗ :N − 3)/(2 ∗ :N) ∗ (LAST :NL) ∗ :X) :NL
END

TO CUM.SUM :L
```

```
IF EMPTYP BUTLAST :L [OUTPUT :L]
LOCAL "CSL
MAKE "CSL CUM.SUM BUTLAST :L
OUTPUT LPUT ((LAST :CSL) + LAST :L) :CSL
END

TO SCALAR.MULT :A :L
IF EMPTYP BUTLAST :L [OUTPUT(LIST :A * (LAST :L))]
OUTPUT LPUT (:A * (LAST :L)) (SCALAR.MULT :A BUTLAST :L)
END
```

The last two programs on this list are dictated by the fact that LOGO does not have the summation of lists and their multiplication by real numbers[5] as primitives. The main part of the overall program is the first pair of programs. I could have done a single program more closely resembling that for the *TI-83*. The division of the main program into the first two of the programs listed is dictated by my penchant for writing recursive procedures and the fact that it is often convenient in such procedures to split them into two programs, one setting things up and one doing the actual recursion.

Of the various methods for finding the square root, Hero's method and the Newton-Raphson Method are the conceptually simplest and most memorable. And the Newton-Raphson Method is very general. If we replace the parabola by any curve, we can draw a similar picture. Starting with any approximation a_0 close to the root, we can generate a sequence of (presumably) better and better approximations $a_1, a_2, a_3, \ldots$ by finding the equation of the line tangent to the curve—something we learn how to do in the Calculus—and then finding where the tangent crosses the x-axis. Provided the curve isn't too flat near the root, the sequence $a_0, a_1, a_2, \ldots$ will converge fairly rapidly. Since Hero's estimate is the same as that given by the Newton-Raphson Method, the same is true of it. So why was the third century Chinese algorithm so popular in the Middle Ages when Hero's simpler, easier to learn method must surely also have been known? The possibly surprising answer is that it is the easiest method short of pushing a button on one's calculator.

Certainly, it is not the easiest of the algorithms to remember. In the others, the steps are guided by theory, but the theory can be a bit advanced. Horner's and Newton's substitution methods require a knowledge of elementary algebra, the Newton-Raphson Method a little analytic geometry, the approach via the Euclidean Algorithm some facility in handling surds and negative numbers, and Newton's Binomial Theorem a bit more algebra. Hero's method seems the simplest of all. A middle school student could readily understand its workings and remember it. But then he would have to use it and come face to face with

---

[5] The name of the last program is dictated by the fact that in vector algebra, the multiplication of a vector by a real number is called *scalar multiplication*. The reason for this is that such multiplication more-or-less scales the given vector up or down in size. Such vectors are often represented as lists of coordinates and the multiplication is carried out by multiplying each of the elements of the list by the given *scalar* as is done here.

the arithmetic. Calculating by hand, this quickly involves some unpleasant long divisions and multiplications with multi-digit numbers. And multiple digits are the Achilles' heel of the calculator's built-in arithmetic operations, as we saw dramatically at the end of our discussion of the value of π.

It is perhaps time we actually programmed our calculators to handle more digits. This is actually quite easy. On reflexion, I realised that I had already done much of the work while writing my *History of Mathematics; A Supplement* (Springer-Verlag, New York, 2008). In working on the chapter therein on Horner's method, I got tired of doing calculations on polynomials by hand and programmed the calculator to do the arithmetic for me. A polynomial, say

$$7X^5 + 4X^4 + 3X^2 - 2X + 1,$$

can be represented as the list,

$$\{7\ 4\ 0\ 3\ -2\ 1\},$$

of its coefficients and the basic arithmetic operations on polynomials can be carried out using the the ordinary arithmetic operations on the entries (since I didn't use coefficients with too many digits). With a couple of minor additional programs, the programs I used there can be used to handle numbers with lots of digits. For, the coefficients of a polynomial $P(X)$ are just the "digits" of P written in base X. And we have but to represent our numbers with lots of digits as lists of digits in some base. For example, the long number 12345678987654321 can be written as

$$\{1\ 2345\ 6789\ 8765\ 4321\}$$

in base 10000 or as

$$\{1\ 23\ 45\ 67\ 89\ 87\ 65\ 43\ 21\}$$

in base 100. And, but for carrying and borrowing, the operations here are the same as those for polynomials with such lists serving as lists of coefficients.[6]

There now follows a longish catalogue of programs for the *TI-83 Plus*. The reader with no calculator or no interest can skip this material. The reader with a *TI-83 Plus* or similar graphing calculator might want to try them out on the calculator, or, at least peruse them for some general insights. They are nice examples of how one can use lists to perform symbolic calculations.

I shall start with programs to perform arithmetic operations on polynomials, and then discuss what needs to be added to perform arithmetic operations on large numbers represented as lists of digits in base 100.

I'm going to use the calculator's built-in addition of lists to add polynomials. To do this, the lists must have the same dimension. We can accomplish this with no change to the polynomial represented by placing some 0's at the front of the shorter list to pad it. This is done by means of the following program:

PROGRAM:PAD
:augment(seq(0,I,1,M),L₁)→L₁

---

[6] Strictly speaking, this is the case for addition, subtraction, and multiplication. Long division is much simpler for polynomials.

That's it. The program assumes one has stored a list in the list variable L_1 and the number of 0's one wishes to prefix the list by in the variable M. The built-in function seq(*expr,var,start,finish*) evaluates an expression using values *start, start*+1, . . . , *finish* for the variable *var* and stores them in order in a list. augment(*list1,list2*) takes two lists as input and creates a longer single list that enumerates the elements of the first list and then follows them with the elements of the second. Thus, if L_1 is {5,1,2} and M is 3, seq(0,I,1,3) is {0,0,0} and augment({0,0,0}, {5,1,2}) is {0,0,0,5,1,2}.

The program PAD is not necessary at all, as it has just one line of code that can be used wherever it is needed. Moreover, the line is functional and functional calls are simpler to make than procedural ones with global variables that have to be assigned values before the calls are made: One simply substitutes the desired values for M and L_1 that one wants to apply the function to. The extra steps involved in calling a procedural program can be simply illustrated by rewriting PAD as a two-step procedure:

```
PROGRAM:PAD
:seq(0,I,1,M)→∟FRONT
:augment(∟FRONT,L₁)→L₁
:DelVar ∟FRONT
```

The first instruction creates a list ∟FRONT of the desired dimension consisting only of 0's, or replaces an already existing list of that name by such a list of 0's. If no list named ∟FRONT already exists, one can achieve the same by using the instruction

```
:M→dim( ∟FRONT)
```

If, however, ∟FRONT already exists, this command merely truncates ∟FRONT if its dimension is larger than M or adds additional 0's at the end of the list if it is too short. To guarantee that all the elements of the list are 0's, one would have to follow the command by

```
:Fill(0,∟FRONT),
```

which replaces all the elements of ∟FRONT by 0's, or

```
:0∗∟FRONT→∟FRONT,
```

which accomplishes the same task by multiplying each entry by 0.

The DelVar command is used to delete variables that are no longer needed, thus freeing up memory, which can be quite limited in the older calculators. It is good practice to add such commands at the ends of one's helper programs to delete the variables one no longer needs in the calling program—but only the unnecessary variables: When in doubt, don't delete.

If one is not aware of the seq(and augment(functions, one could program the steps they take. Thus, PAD would look like

```
PROGRAM:PAD
:dim(L₁)→ N
:M+N→dim(∟TEMP)
```

```
:For(I,1,M)
:0 →∟TEMP(I)
:End
:For(I,1,N)
:L₁(I)→∟TEMP(M+I)
:End
:∟TEMP→L₁
:DelVar ∟TEMP
:DelVar N
:DelVar I
```

Here, the assignment :M+N→dim(∟TEMP) gives us a list ∟TEMP of length M+N, the first M values of which we sequentially overwrite with 0's in the first For loop and into the last N entries of which we copy L_1 using the second For loop. We then copy ∟TEMP into L_1 and delete the variable ∟TEMP which has served its purpose. We also delete the variables N and I which currently have the dimension of the original L_1 stored in them—information which is no longer needed.

Accompanying PAD, we need a program UNPAD to remove excess 0's from the front of a list. Such a program will naturally begin by counting the number of 0's there are at the beginning of a list, which we assume to be stored in L_1.

```
PROGRAM:COUNTER
:dim(L₁)→ N
:0 → K
:While L₁(K+1) = 0
:K+1→ K
:If K = N
:Return
:End
```

The Return command ends the execution of the program, returning to where it was called if COUNTER was executed as the result of a program call in another program. End commands are used to indicate the ends of the scopes of For, While, If... Then, and If... Then... Else commands. The final :End here finishes off the While loop. [The If command, unaccompanied by Then, does not need an End. Only the immediately following instruction (in this case, Return) is executed and this only if the condition cited is true.]

Running the COUNTER program results in the number of 0's at the beginning of L_1 being stored in the variable K and the dimension of L_1 stored in N. We need both of these values in UNPAD:

```
PROGRAM:UNPAD
:prgmCOUNTER
:If N = K
:Then
:{0} → L₁
:Else
```

```
:seq(L₁(I),I,K+1,N)→L₁
:End
:DelVar N
:DelVar K
```

The general strategy of the program is simply to copy all but the first K 0's of L_1 into a new version of L_1. The only problem is that, if all entries are 0, there is nothing to copy and, not only do we not want the empty list, but the calculator doesn't allow it. Thus, in the case $K = N$, i.e., L_1 is $\{0,0,\ldots,0\}$, we simply overwrite L_1 with $\{0\}$. [Note that the calculator is of two minds concerning lists: You enter a list using curly braces $\{,\}$, separating the elements with commas, but the answer returned on the screen with the elements separated by spaces.]

5 Exercise. Write a program REVERSE which takes the list L_1 and reverses the order of its elements. Write a new version of UNPAD replacing the line

```
:seq(L₁(I),I,K+1,N)→L₁
```

by the lines

```
:prgrmREVERSE
:N−K→dim(L₁)
:prgmREVERSE
```

Copy the new UNPAD into your calculator and try it on the lists $\{0\}$, $\{0,0,5,2\}$, $\{5,0,2\}$.

The PAD and UNPAD programs are just helper programs. The actual arithmetic is carried out by programs POLYSUM, POLYDIFF, POLYMULT, and POLYDIV. The first two of these are:

```
PROGRAM:POLYSUM              PROGRAM:POLYDIFF
:dim(∟POLY1)→A               :dim(∟POLY1)→A
:dim(∟POLY2)→B               :dim(∟POLY2)→B
:If A < B                    :If A < B
:Then                        :Then
:B−A→ M                      :B−A→ M
:POLY1→L₁                    :POLY1→L₁
:prgmPAD                     :prgmPAD
:L₁→∟POLY1                   :L₁→∟POLY1
:End                         :End
:If B < A                    :If B < A
:Then                        :Then
:A−B→M                       :A−B→M
:POLY2→L₁                    :POLY2→L₁
:prgmPAD                     :prgmPAD
:L₁→∟POLY2                   :L₁→∟POLY2
:End                         :End
```

```
:∟POLY1 + ∟POLY2→L₁          :∟POLY1 − ∟POLY2→L₁
:prgmUNPAD                    :prgmUNPAD
:L₁→∟PSUM                     :L₁→∟PDIFF
:ClrList L₁                   :ClrList L₁
:DelVar A                     :DelVar A
:DelVar B                     :DelVar B
:DelVar M                     :DelVar M
```

These programs are rather long, but simple. First one finds the lengths of the lists, stores the difference in M, and pads the shorter of the two lists. One then uses the built-in addition (respectively, subtraction) of lists to carry out the actual calculation. One then unpads the result and stores it in a list ∟PSUM or ∟PDIFF, respectively. The real workhorses are the single lines ∟POLY1 + ∟POLY2→L₁ and ∟POLY1 − ∟POLY2→L₁; everything before them is merely preparing the data and everything after is cleaning up.

Note too that, once one has defined POLYSUM, one can give a shorter subtraction program:

```
PROGRAM:POLYDIFF
:−∟POLY2→∟POLY2
:prgmPOLYSUM
:∟PSUM→∟PDIFF
:DelVar ∟PSUM
```

Being substantially shorter, it takes up less space in RAM. Its execution, on the other hand, requires a few more steps: negating all the elements of ∟POLY2, copying ∟PSUM into ∟PDIFF, and deleting a redundant variable. It is a trade-off, the decision favouring the longer version if POLYDIFF is going to be called a large number of times. (Another thing: Should one repeat the first instruction to restore ∟POLY2, or should one delete it and ∟POLY1? The answer depends on whether one is finished with these polynomials in the larger context.)

The programs for multiplication and division are

```
PROGRAM:POLYMULT              PROGRAM:POLYDIV
:POLY1→L₁                     :POLY1→L₁
:prgmUNPAD                    :prgmUNPAD
:L₁→∟POLY1                    :L₁→∟POLY1
:POLY2→L₁                     :POLY2→L₁
:prgmUNPAD                    :prgmUNPAD
:L₁→∟POLY2                    :L₁→∟POLY2
:dim(∟POLY1)→K                :dim(∟POLY1)→M
:dim(∟POLY2)→M                :dim(∟POLY2)→N
:K+M−1→N                      :If M >N
:seq(0,I,1,N)→∟PPROD          :Then
:For(I,1,K)                   :{0} →∟PQUOT
:For(J,1,M)                   :∟POLY2→∟PREM
```

```
:∟POLY1(I)∗∟POLY2(J)                    :Else
 +∟PPROD(I+J−1)→∟PPROD(I+J−1)          :N−M→ K
:End                                    :K+1→dim(∟PQUOT)
:End                                    :∟POLY1→ L₂
:DelVar K                               :N→dim(L₂)
:DelVar M                               :For(I,1,K)
:DelVar N                               :L₁(I)/∟POLY1(1)→ C
:DelVar I                               :C→∟PQUOT(I)
:DelVar J                               :C∗L₂→ L₃
:ClrList L₁                             :L₁−L₃→ L₁
                                        :augment({0},L₂) → L₂
                                        :N →dim(L₂)
                                        :End
                                        :L₁(K+1)/∟POLY1(1) → C
                                        :C→∟PQUOT(K+1)
                                        :C∗L₂→L₃
                                        :L₁−L₃→ L₁
                                        :prgmUNPAD
                                        :L₁→∟PREM
                                        :End
                                        :DelVar M
                                        :DelVar N
                                        :DelVar K
                                        :DelVar I
                                        :DelVar C
                                        :ClrList L₁
                                        :ClrList L₂
                                        :ClrList L₃
```

In POLYSUM and POLYDIFF, we started the programs by padding one of the lists so that the dimensions matched and we could apply the built-in + and − operations. The dimensions do not need to match when multiplying and dividing polynomials, so we prefix no padding. In fact, I have begun each program by unpadding the lists. For POLYMULT this is unnecessary, but may cut out a lot of unnecessary multiplications by 0 and additions of 0. And, in POLYDIV, where we divide ∟POLY2 by ∟POLY1, unpadding ∟POLY1 is necessary to avoid dividing by 0 in the lines :$L_1(I)/$∟$POLY1(1)\to$ C and :$L_1(K+1)/$∟$POLY1(1)\to$ C. Unpadding ∟POLY2 is again merely a convenience.

After the unpadding operations, the next order of business in POLYMULT is to create a list ∟PPROD that will serve as the product. If the dimensions of ∟POLY1 and ∟POLY2 are K and M, respectively, their respective degrees are $K − 1$ and $M − 1$, whence their product will be of degree $K − 1 + M − 1$ $= K + M − 2$ and the list will therefore have dimension $K + M − 1$. This is stored in N and ∟PPROD is created as a list of N 0's. Then the coefficients of the individual product terms of degrees 0 to N − 1 are added one at a time to the appropriate elements of ∟PPROD until the final product results. This program doesn't exploit the built-in calculator operations the same way that POLYSUM and POLYDIFF did. One can do this by simulating the usual cascade of multidigit multiplication by storing the results of the cascade in a matrix. Thus, e.g., for

$$
\begin{array}{r}
2\ 2 \\
3\ 1 \\
\hline
2\ 2 \\
6\ 6 \\
\hline
6\ 8\ 2
\end{array}
$$

we can store the part between the parallel lines in a matrix

$$[A]: \begin{bmatrix} 0 & 2 & 2 \\ 6 & 6 & 0 \end{bmatrix}$$

We can then add the columns using the matrix operation

$$\mathsf{cumSum}([A]) \rightarrow [A]$$

resulting in

$$\begin{bmatrix} 0 & 2 & 2 \\ 6 & 8 & 2 \end{bmatrix}$$

and then reading off the bottom row as a list,

$$\mathsf{seq}([A](2,\mathsf{I}),\mathsf{I},1,3) \rightarrow \text{∟PPROD}.$$

Or, we can use matrix multiplication to get the row wanted:

$$
\begin{bmatrix} 0 & 2 & 2 \\ 6 & 6 & 0 \end{bmatrix}^{\top}
\begin{bmatrix} 1 \\ 1 \end{bmatrix}
=
\begin{bmatrix} 0 & 6 \\ 2 & 6 \\ 2 & 0 \end{bmatrix}
\begin{bmatrix} 1 \\ 1 \end{bmatrix}
=
\begin{bmatrix} 0+6 \\ 2+6 \\ 2+0 \end{bmatrix}
=
\begin{bmatrix} 6 \\ 8 \\ 2 \end{bmatrix}
$$

Of course, the matrix on the right, say [B], is a matrix, not a list, but there is a built-in matrix-to-list operation that can be exploited for this purpose:

$$\mathsf{Matr} \blacktriangleright \mathsf{list}(\,[B],\text{∟PPROD}).$$

This approach, using as many built-in functions as possible, is more in line with our treatment of POLYSUM and POLYDIFF, but it smacks of striving for cleverness when POLYMULT is so much simpler conceptually. It may, however, result in a faster program.

I won't attempt an explanation of POLYDIV other than to remark that it divides a polynomial ∟POLY2 by another polynomial ∟POLY1, resulting in a quotient ∟PQUOT and remainder ∟PREM, the latter being of degree less than that of ∟POLY1. Working through the steps of the program by hand for one or two specific examples will explain it better than words can. In any event, POLYDIV is just long division for polynomials and is much, much simpler than ordinary long division with numbers, which we will not be needing in discussing square roots. The important programs for us are POLYSUM, POLYDIFF, and POLYMULT. The only thing that keeps us from using them to perform arithmetic operations on large numbers presented as lists of digits in some given base is the fact that polynomials allow large coefficients as well as negative ones: no carrying or borrowing operations are performed by these programs.

If, say,

$$\{6, 2, 1\} \text{ and } \{5, 3, 4\}$$

represent quadratic polynomials $6X^2 + 2X + 1$ and $5X^2 + 3X + 4$, respectively, and we assign these lists as values to the lists ∟POLY1 and ∟POLY2, respectively, then running POLYSUM and POLYDIFF results in $\{11,5,5\}$ and $\{1,-1,-3\}$, respectively, and multiplying ∟POLY1 by 2 results in $\{12,4,2\}$. These are absolutely correct if the lists are taken to represent polynomials, but not if the elements of the lists are taken to be digits of numbers written in base 10. Numerically, they are correct; e.g.,

$$11 \cdot 100 + 5 \cdot 10 + 5 = 10 \cdot 100 + 1 \cdot 100 + 5 \cdot 10 + 5$$
$$= 1 \cdot 1000 + 1 \cdot 100 + 5 \cdot 10 + 5$$
$$= 1155 = 621 + 534,$$

and

$$1 \cdot 100 - 1 \cdot 10 - 3 \cdot 1 = 1 \cdot 100 - 2 \cdot 10 + 1 \cdot 10 - 3 \cdot 1$$
$$= 1 \cdot 100 - 2 \cdot 10 + 7 \cdot 1$$
$$= 0 \cdot 100 + 10 \cdot 10 - 2 \cdot 10 + 7 \cdot 1$$
$$= 8 \cdot 10 + 7 \cdot 1 = 87 = 631 - 534.$$

All we have to do is perform the indicated carrying and borrowing operations to find a list representing the same number but with integral digits in the range 0 - 9.

For the square root algorithm, numbers are written in base 100. These operations are, of course, performed from right to left. For carrying, when one reaches the leftmost entry, if a carrying operation needs to be performed, one has to prefix the list with a new element. And with borrowing, one may have to unpad the list if the leftmost element has been reduced to 0.

Without further ado, here are the CARRY and BORROW programs for base 100.

PROGRAM:CARRY
:dim(L$_1$)→ N
:For(I,0,N−1)
:If L$_1$(N−I) ≥ 100
:Then

PROGRAM:BORROW
:dim(L$_1$)→ N
:If N > 1
:Then
:For(I,0,N−2)

:int(L$_1$(N−I)/100)→Q   :If L$_1$(N− I) < 0
:L$_1$(N−I)−100*Q→L$_1$(N−I)   :Then
:If I = N−1   :L$_1$(N−I)+100→L$_1$(N−I)
:Then   :L$_1$(N−I−1)−1→L$_1$(N−I−1)
:augment({Q},L$_1$)→L$_1$:End
:Else   :End
:L$_1$(N−I−1) + Q→L$_1$(N−I−1) :prgmUNPAD
:End   :End
:End
:End
:DelVar N
:DelVar I
:DelVar Q

A couple of quick comments: For the addition of two positive numbers in normal form, Q will always be 1 and the program can be simplified ever so slightly. But, even multiplying by a single digit number[7] could result in a larger number being carried. For example,

$$99 * \{1, 90, 77, 2\}$$

yields

$$\{99\ 8910\ 7623\ 198\}$$

and one will have to carry a 1, then a 76, then 89, and then a 1, resulting in $\{1\ 88\ 86\ 24\ 98\}$. A second remark concerns **BORROW**: I have not included commands deleting N and I. These are global variables and their values are changed automatically during execution of **UNPAD**, and at the end of this program's execution the variables are deleted.

We have enough machinery on hand to accomplish some tasks with the calculator that on first sight might seem beyond its power.

6 Example. If you applied the polynomials P and Q defined at the end of the chapter on the Pell equation to -1, you should have got

$$x = 1766329049, \quad y = 226153980$$

as a solution to the Bhāskara-Fermat-Euler instance,

$$X^2 - 61Y^2 = 1,$$

of the Pell equation. Now it may take a while, but you can check that this is indeed a solution by squaring x and y by hand, multiplying y^2 by 61 and subtracting the result from x^2. But if you decide to spare the effort you might

---

[7] "Single digit" relative to the given base. Here, of course, I could have multiplied by 9 and still have had a couple of carries to perform with digits greater than 1.

want to try entering x, y into the variables X,Y on the calculator, and entering the expression

$$X^2 - 61Y^2.$$

The result returned is 0, which is simply wrong as we can see effortlessly by noting that x is odd, y is even, whence $x^2 - 61y^2$ is odd and cannot be 0. What we can do is represent x as the list $\{17,66,31,90,49\}$ and y as $\{2,26,15,39,80\}$ and use the programs given, perhaps storing the lists in list variables ᴸX, ᴸY, respectively, storing the number 61 in a variable D, and then writing a simple master program:

```
PROGRAM:CHECK
:ᴸX→ᴸPOLY1
:ᴸX→ᴸPOLY2
:prgmPOLYMULT
:ᴸPPROD→L₁
:prgmCARRY
:L₁→ᴸX2
:ᴸY→ᴸPOLY1
:ᴸY→ᴸPOLY2
:prgmPOLYMULT
:ᴸPPROD→L₁
:prgmCARRY
:D∗L₁→L₁
:prgmCARRY
:L₁→ᴸPOLY2
:ᴸX2→ᴸPOLY1
:prgmPOLYDIFF
:ᴸPDIFF→L₁
:prgmBORROW
:prgmUNPAD
```

The excessive length of the program (which does not yet include all the DelVar commands) is dictated by the fact that we cannot define functions by programs and must use global variables. Most of the operations given are just copying lists into different variables, either to apply a program to them, or to store them elsewhere temporarily while the variable is overwritten by another list to which we wish to apply some program that uses the given variable.

If one now runs the program, it takes a while (several seconds), but it produces the list $\{1\}$ in L₁, verifying that $x^2 - 61y^2 = 1$. Again, using another value of D proposed by Fermat, store 109 into D, $\{1,58,7,6,71,98,62,49\}$ into ᴸX, and $\{15,14,4,24,45,51,0\}$ into ᴸY and run CHECK. This will result in L₁ being $\{1\}$. The program takes a number of seconds and I haven't tried it on the solution for D = 189574 given by the numerator and denominator of the fraction (5), but, if the calculator doesn't run out of memory, it should take quite a bit of time—you might start the program and settle down with a nice book.[8] (Generally, if I want to do something this computationally intensive, I

_____

[8] Perhaps I exaggerate.

do it on a computer in SCHEME, which is quite fast. The verification, indeed the generation of the solution for $D = 189574$, which included the check as part of the procedure, took less than a second on my *iMac* running SCHEME.)

Let us now get down to programming the square root algorithm with which we opened the present chapter. We have all the necessary tools to program our *TI-83 Plus* to carry out the algorithm to as many steps as we choose. Indeed, we have more than we need as we will only be multiplying by single digit numbers $0, 1, \ldots, 9$, and for this we do not need the program POLYMULT but can use the calculator's built-in multiplication of lists by numbers. What we need are PAD, UNPAD, POLYDIFF, CARRY, and BORROW.

Before we begin programming, we have to think about how we are going to represent data in the calculator. First, to simplify matters, I propose to find the square roots of whole numbers only. Handling decimals can be done by moving the decimal point before and after the program has run. We want to use d as a number in base 100, so we could assume d to have been entered into the calculator as a list ∟D of base 100 digits. If d itself has 10 or fewer digits in base 10, we could enter it as a number and convert it to a list in base 100 via a program like the following:

```
PROGRAM:CONVNUM :D→N
:int(log(N))→M
:{100*fPart(N/100)}→∟D
:iPart(N/100)→N
:For(I,1,int(M/2))
:augment({100*fPart(N/100)},∟D)→∟D
:iPart(N/100)→N
:End
:DelVar M
:DelVar N
```

If d has more than 10 digits, it has to be entered into the calculator as a list, say

$\{1,2,3,4,5,6,7,8,9,8,7,6,5,4,3,2,1\}$→$L_1$,

or as a string,

"12345678987654321"→Str1

One would then write a conversion program CONVLST or CONVSTR, respectively. Entering the number as a string is perhaps more convenient as one doesn't have to enter all those commas. The conversion program for strings goes

```
PROGRAM:CONVSTR
:length(Str3)→M
:0→dim(L₁)
:If gcd(M,2)=1    (i.e., if M is odd)
:Then
```

```
:expr(sub(Str3,1,1))→L₁(1)
:For(I,1,(M−1)/2)
:expr(sub(Str3,2I,2))→L₁(I+1)
:End
:Else
:For(I,1,M/2)
:expr(sub(Str3,2I−1,2))→L₁(I)
:End
:End
:DelVar M
:DelVar I
```

A few words of explanation are in order. The calculator has only 10 string variables Str1, ..., Str9, Str0. I wish to reserve Str1 and Str2 for two special strings and thus use Str3 as a working variable. The program will be called twice from the master program, to convert both Str1 and Str2 into special lists. Thus, I use L_1 here instead of ∟D as in CONVNUM.

I also note that the functions length(, expr(, and sub(are accessed only through the CATALOG, and the string variables are accessed via the VARS menu. length(gives the length of a string, expr(evaluates the string. Note that the string "11" and the number 11 are separate entities. expr("11") evaluates to the number 11. sub(Str$n$,$start$,$length$) returns the substring of the string Str$n$ of length $length$ beginning at position $start$ of Str$n$.

7 Exercise. Write the program CONVLST.

Just as we can represent d in various ways, there is more than one way of representing the square root of d. If there are 10 or fewer digits, we can represent it as a number. If there are more than 10, a list or a string is called for. As the algorithm generates one base 10 digit at a time, the natural list representation would be a list of base 10 digits. The syntax is slightly different with the two choices, but the overall program structure isn't. Strings have a slight advantage in that, as they can include any characters, we can incorporate the decimal point in the final representation of the square root. Against this, strings are slightly clumsier to deal with than lists. However, overcoming this clumsiness affords us some useful practice in dealing with strings. Thus, I shall choose to represent the square root as the string Str2.

The square root algorithm will continue forever until we arbitrarily declare an end to it. In our program, the stopping point will be decided in advance by our storing the desired number of digits to the right of the decimal point in the global variable P (for $precision$). Rather than have to remember which variable is which, we might write our program as an interactive one that asks us to enter d as a string and a value for P. It could start something like

```
PROGRAM:SQROOT
:Disp "ENTER A"
:Disp "WHOLE NUMBER,"
:Input "D=", Str1
```

```
:Str1→Str3
:prgmCONVSTR
:L₁→∟D
:ClrList L₁
:DelVar Str3
:Disp "HOW MANY"
:Disp "DECIMALS"
:Disp "ACCURACY?"
:Input "P=", P
```

A full-blown interactive program would test Str1 to make sure the string indeed consists solely of digits and would loop back if it wasn't (i.e., if it contained characters other than $0, 1, \ldots, 9$). Likewise, it would test to make sure P was a whole number. Mine will be a bare bones program that assumes the user enters acceptable values for these variables. The manual for the *TI-83 Plus* does not say how long a string can be, but I've stored strings a couple thousand characters long on my calculator, so I assume it stores sufficiently many for our purpose. I also assume we are not seriously interested in finding many millions of digits and that P will be small enough (say, 2 or 3 digits) that we can enter it as an ordinary number.

The length of ∟D tells us how many digits will lie to the left of the decimal point in $\sqrt{d}$ and P tells us how many will be presented to the right. Our first computational steps will be to store the former length into a new variable L and to move the centesimal point of ∟D P places to the right:

```
:dim(∟D)→L
:L+P→C
:C→dim(∟D)
```

Global variables offer a challenge to the programmer. If we have a quantity that has to be preserved throughout, we don't want to store it in a variable that is going to be changed in a called program. The programs we have listed so far that we will be calling are PAD, UNPAD, POLYDIFF, CARRY, BORROW (and thus COUNTER), and CONVSTR. They use the number variables A, B, I, K, M, N, Q, the list variables L₁, ∟TEMP, ∟POLY1, ∟POLY2, ∟PDIFF (and possibly ∟FRONT, depending on the version of PAD used), and the string variable Str3. To these we now add C, L, P, ∟D, and Str1, all of which we will want not to change. And I have already declared Str2 reserved for the string of digits of $\sqrt{d}$. The more generous variable naming capabilities of a calculator like the *TI-85* would make writing a readable program a lot easier. And the ability to program functions on the *TI-89* should simplify matters even further.

Looking at FIGURE 1, we see that there are three quantities that vary over time: the initial a and succeeding b's (4, 3, 5, ...), the minuend (18, 295, 4674, 34900, ...), and the subtrahend (16, 249, 4325, 34810, ...). The algorithm proceeds by replacing each in turn by its new value.

Our initial a and succeeding b's cannot very well be called A and B, which global variables have already been used. As they are successive digits of the desired square root, we may as well call them S for square root. The other two lists (18, 295, 4674, ... and 16, 249, 4325, ... in FIGURE 1) each begins with a number and thereafter consists of numbers we will be thinking of as numbers written in base 100, thus presented as lists. As the main thing we do with them is subtraction, we might as well call them ∟POLY1 and ∟POLY2, respectively.

The first step in the algorithm is estimating the square root of the first base 100 digit of ∟D:

```
:int(√(D(1)))→S
:prgmMAKESTR
:Str3→Str2
:DelVar Str3
:{∟D(1)−S²}→∟PDIFF
```

The program MAKESTR puts the value of the variable S into a string Str3. I offer it in two versions. The first is a stupid-looking program, but it is instantly intelligible and it does the trick. The second I found online at the Texas Instruments web site. It is shorter and works for multi-digit numbers, but is devious and I refrain from discussing it.

```
PROGRAM:MAKESTR          PROGRAM:MAKESTR
:If S=0                  :{1,2}→L₃
:"0"→Str3                :{S,S}→L₄
:If S=1                  :LinReg(ax+b) L₃,L₄,Y₀
:"1"→Str3                :Equ▶String(Y₀,Str3)
:If S=2                  :length(Str3)→L
:"2"→Str3                :sub(Str3,4,L−3)→Str3
:If S=3                  :ClrList L₃,L₄
:"3"→Str3                :DelVar L
:If S=4                  :DelVar Y₀
:"4"→Str3
:If S=5
:"5"→Str3
:If S=6
:"6"→Str3
:If S=7
:"7"→Str3
:If S=8
:"8"→Str3
:If S=9
:"9"→Str3
```

I have separated these lines into a distinct program because SQROOT will have to call it again to extend Str2 (whence the need for a new variable Str3).

Most of the rest of the program SQROOT is a loop:

```
:For(J,2,C)
:augment(∟PDIFF,{∟D(J)})→∟POLY1
```

```
:POLY1→L₁
:∟program UNPAD
:L₁→POLY1
:prgmGETVARS
:prgmMAKESTR
:Str2+Str3→Str2
:prgmPOLYDIFF
:∟PDIFF→L₁
:prgmBORROW
:prgmUNPAD
:L₁→∟PDIFF
:End
```

Obviously this needs some explanation. A simple matter is the choice of the variable J as a counter. The usual choices, I, K, N are used in the programs PAD, UNPAD and BORROW, the first two being called by POLYDIFF. J is not used in any of these programs and can safely be used here with no fear of its being overwritten. ∟POLY1 is obtained by appending the next centesimal digit of ∟D to the difference, ∟PDIFF, obtained in the previous cycle of the algorithm. Hence the first line in the loop. The variables S and ∟POLY2 are more complicated and will be supplied by a separate program cleverly named GETVARS. Once this is run, S and ∟POLY2 have been defined, MAKESTR turns S into a string Str3, which is usd to extend Str2 by concatenation, which for strings is performed on the *TI-83 Plus* by using the + button. All that remains now is to subtract ∟POLY2 from ∟POLY1 and close the loop.

The crucial thing now is to describe **GETVARS**. To get the variables, one must go back to FIGURE 1 and the working of the algorithm. The approximation to $\sqrt{d}$ going into a loop is given by the current value of Str2 and d is approximated by an initial segment. We bring down the next centesimal digit of d and multiply Str2 by 10 and take the difference $d - (10a)^2$ with the new values of d and a. This turns out to be the number represented by ∟POLY1. (In FIGURE 1, $18 - 4^2 = 2, 1895 - 40^2 = 295, 189574 - 430^2 = 4674$, ec.) So we have to find the largest b such that $(20a + b)b <$ ∟POLY1. Previously, we simply divided ∟POLY1 by $20a$ and concluded b was the greatest integer in the fraction $b - 1$ or $b - 2$ or.... Since b is limited to being one of $0, 1, \ldots, 9$, it is easier to write a program testing these values in turn until b is found. We then store b in S and $(20a + b)b$ in ∟POLY2. To do this, we need a comparison program to decide when a list L_2 is less than L_1:

```
PROGRAM:LESS
:0→T
:dim(L₁)→A
:dim(L₂)→B
:If A<B
:Goto 1
:If A>B
:Then
```

```
:1→T
:Goto 1
:End
:For(I,1,A)
:If L₁(I)<L₂(I)
:Goto 1
:If L₁(I)>L₂(I)
:Then
:1→T
:Goto 1
:End
:End
:Lbl 1
:DelVar A
:DelVar B
:DelVar I
```

LESS assumes L_1 and L_2 are lists representing numbers in base 100 in normal form (i.e., no leading 0's, all entries integers from 0 to 99). It starts by storing 0 (for "false") in a new variable T. If L_2 has fewer centesimal digits than L_1 or if the number of digits is the same and $L_2(i) < L_1(i)$ at the first place they differ, then 1 (for "true") replaces 0 in T; otherwise T is left equal to 0. The end result is that T contains a 1 iff the number represented by L_2 is less than that represented by L_1.

```
PROGRAM:GETVARS
:Str2→Str3
:prgmCONVSTR
:20*L₁→L₁
:prgmCARRY
:dim(L₁)→E
:L₁→L₃
:For(F,0,9)
:L₃→L₁
:L₁(E)+F+1→L₁(E)
:(F+1)*L₁→L₁
:prgmCARRY
:L₁→L₂
:LPOLY1→L₁
:prgmLESS
:If T=0
:Then
:F→S
:L₃→L₁
:L₁(E)+F→L₁(E)
:F*L₁→L₁
:prgmCARRY
```

:$L_1 \to {}_{\llcorner}$POLY2
:Goto 1
:End
:End
:Lbl 1
:DelVar E
:DelVar F
:DelVar T
:DelVar Str3
:ClrList L_1,L_2,L_3

All that remains to complete SQROOT is to put the decimal where it belongs and to do some cleaning up:

:sub(Str2,1,L)+"."+sub(Str2,L+1,P)$\to$Str2
:DelVar C
:DelVar J
:DelVar L
:DelVar P
:DelVar S
:DelVar ${}_{\llcorner}$POLY1
:DelVar ${}_{\llcorner}$POLY2
:DelVar ${}_{\llcorner}$PDIFF

This is not the most efficient program, and the *TI-83 Plus* is not the fastest machine. I started the program, entered 2 for D and 50 for P and from that point it took a few seconds short of 15 minutes for the calculator to come up with

$$1.4142135623730950488016887242096980785696718753769\mathbf{4}. \qquad (11)$$

After all that work it would be nice to know that this is correct. Checking this actually took very little time. First, I erased the decimal point:

"1"+sub(Str2,3,50)$\to$Str3

Then I ran CONVSTR to produce a list

$$\{1, 41, 42, \ldots, 94\}$$

and stored it in ${}_{\llcorner}$POLY1 and ${}_{\llcorner}$POLY2. The program POLYMULT then produced a list which CARRY then transformed to

$$\{1, \underbrace{99, \ldots, 99}_{24}, 97, 71, \ldots\}.$$

I then replaced the final 4 of (11) by 5,[9]

---

[9] I could as well have gone back to Str2:

"1"+sub(Str2,3,49)+"5"$\to$Str3,

or I could even have bypassed Str2 by entering

sub(Str3,1,50)+"5"→Str3,

ran CONVSTR to produce a list

$$\{1, 41, 42, \ldots, 95\},$$

and stored this in ∟POLY1 and ∟POLY2. Running POLYMULT and CARRY produces

$$\{2, \underbrace{0, \ldots, 0}_{25}, 54, 49, \ldots\}.$$

Thus we see that the answer given is indeed correct to the 50th decimal place. (*Exercise.* Why does this follow from the two computations?)

Obviously, there is not much classroom application of all of this in the middle school. If finding square roots by hand makes a comeback in elementary education, having the program will allow you to check your student's answers should you ask them for results of higher accuracy than the calculator yields. That's about as directly useful it can be. But it is kind of fun to do such things and I shall close this chapter with a few projects to think about.

8 Project. How can you make the program more efficient? One example: SQROOT only performs a subtraction when ∟POLY2 represents a smaller number than ∟POLY1. Thus the test to see if A (= the dimension of ∟POLY1) is less than B (= the dimension of ∟POLY2) will always result in 0 (for "false") and that step can be cut out. Are there any other built-in inefficiencies in my programs? Also, all the subprograms can be incorporated into a single program and then it can be compiled as an assembly language program. Assembly language programs run faster.

9 Project. Long division, like taking square roots, can go on forever. Write a program LONGDIV which, given two numbers m, n and a level p of precision, will find m/n to p decimal places.

10 Project. (Assumes Calculus). In Chapter 1, it was mentioned that al-Kāshī used a polygon with $6 \cdot 2^{27}$ sides to calculate π to 16 decimals. What level of accuracy in calculating square roots is necessary to achieve that accuracy by the method described? (Note that it requires a nesting of 28 square roots.)

11 Project. Suppose you want to make a table of square roots of all the integers from 1 to 100 to 50 decimal places. Ignoring time spent copying the answers from your calculator, how many hours would you estimate it would take? Can you shave $2\frac{1}{2}$ hours off this right away? (Or, have you already done this in making your initial estimate?) Multiplying two 50+ digit numbers takes a lot less time than finding a single square root to 50 decimal places. If estimates for $\sqrt{a}, \sqrt{b}$ are correct to 50 places, how well does their product estimate $\sqrt{ab}$?

95→∟POLY1(26)   and   ∟POLY1→∟POLY2.

12 Project. The verification that (11) is indeed the square root of 2 suggests a simpler program for generating square roots:

> Given a value d of D, let ∟DTEMP be the first or first two digits of d as the number of digits of d is odd or even, and let ∟RT be the list whose only element is the greatest integer less than or equal to the number represented by ∟DTEMP. At each successive step append the next two digits of d to ∟DTEMP and find the largest k such that appending k to ∟RT results in a list whose square (as calculated by POLYMULT is less than ∟DTEMP. Append k to ∟RT as the next digit in the root, and continue.

Write such a program for the *TI-83 Plus* and use it to find the square root of 2 to 50 decimal places. How does the program compare to SQROOT? in terms of ease of programming and efficiency?[10] Which algorithm would you prefer to use if you were a mediæval European mathematician calculating square roots to, say, 14 or 15 decimal places?

13 Project. Look up some of the old algorithms for finding cube roots and proceed from there.

---

[10] For me, the new program was much easier to write. I believe it was about as efficient as could be expected for the algorithm described. After half an hour only 25 decimals had been calculated, the next fifteen minutes added only 5 more digits, and each of the next three quarter hours brought only 2 digits apiece. I cannot report how long it takes to obtain all 50 digits because I was working on something else while the program was running and at the 1 hour 40 minute mark I absent-mindedly picked up my calculator and started punching buttons, interrupting the program's execution. Thus I aborted the program and checked the results up to that point. All 36 digits following the decimal point were correct, so I was pleased with that. It occurred to me that with only 2 digits per 15 minutes, it would take at least another hour and 45 minutes to obtain the last 14 digits of $\sqrt{2}$, thus over 3 hours in all and I just did not feel like wasting the time running the experiment again.

Using the Computer: Square Roots and the Pell Equation Revisited

[The material of this chapter strays a bit from our main path and can safely be omitted. I had fun writing and testing the programs and thought I would report on it here for the entertainment of the more adventurous reader.]

Our long discussion of the square root program and all of its auxiliary programs demonstrated how to overcome some of the limitations of the *TI-83 Plus*. It also served, of course, to illustrate what these limitations were. The *TI-89* represents a whole new generation of calculator. It performs symbolic calculations far more complex than the simple arithmetic operations on polynomials given by our POLYSUM, POLYDIFF, ... suite of programs. It handles integers exactly, meaning one wouldn't have to use lists and strings, symbolic computation via POLYDIFF, or the CARRY or BORROW programs in programming the square root algorithm. As with computer languages which allow the exact treatment of large integers, the *TI-89* comes equipped with an implementation of the Division Algorithm—it has two functions for obtaining the quotient and remainder when dividing two positive integers. This means that the For loop of our GETVARS program can be replaced by the simple calculation of the quotient b of

$$\frac{d - a^2}{2a}$$

and the test to see if

$$(2a + b)b < d - a^2$$

to determine whether one takes b or $b - 1$ as the next digit. Moreover, the *TI-89* allows the programming of functions with local variables, thus allowing simpler, more readable programs. To all of this one should add that the *TI-89* is faster and has more memory than the *TI-83 Plus*, whose slow speed was probably the most disappointing feature of our computation of $\sqrt{2}$.

The downside is that the *TI-89* reportedly has a less intuitive interface than the *TI-83 Plus* or its *TI-84* successors, and a much steeper learning curve. The extra power is generally only needed at the university level and beyond, not in high school mathematics and certainly not in the middle school. For this reason I have not considered justified the expense of purchasing a new calculator for the purpose of comparing the execution of a reworked SQROOT program on the *TI-89* with that on the *TI-83 Plus*. I could use the manual to rewrite the program, but I am reluctant to do so without having the calculator to test such a program on.

What I can do is present the modified program for a computer. Powerful programming languages with all the desired features can be downloaded for free for all three major platforms: *Mac, Linux,* and *Windows*. Ideally, I would use LOGO. It is a very powerful and easy to use learning language for elementary programming and it used to be taught at the elementary school level. Unfortunately, it seems to have fallen out of fashion, but I think for no good reason. I would urge every middle school teacher to acquire a copy and learn it. Freeware versions and good supporting documentation are available online. I suggest UCB LOGO from the University of California at Berkeley's website.[1]

Having recommended LOGO, I must admit that it is not my choice for the present chapter. Implementations of LOGO do not treat integers exactly and the programs would overall be the same as those for the *TI-83 Plus*. LOGO does have some conveniences—local variables and the ability to program functions will simplify some of the programs. And the use of a higher speed computer should make a noticeable improvement in performance. But LOGO programs are generally *interpreted*, not *compiled*. This means that the translation from LOGO into the machine's language is done "on the fly". Each line of code is read, translated, and executed, one after another. If one loops an instruction 45 times, the translation will be made 45 times. With many other languages, programs are *compiled*—translated once directly into the machine's language and it is the machine language program that is run when the program is executed. Compiled programs are much faster than interpreted ones.

The language I wish to use is SCHEME. It is similar in structure to LOGO, but a bit less convenient. In LOGO, for example, one can write algebraic expressions like

$$x * x - d * y * y$$

and assume LOGO knows the order of precedence for the operations. In SCHEME one must write

$$(- (* \ x \ x) \ (* \ d \ (* \ y \ y))),$$

or else write one's own interpreter for algebraic expressions. Moreover, the syntax is not based on mathematical usage. For example, the greatest integer less than or equal to a number (represented by int(in the NUM submenu of the MATH menu on the *TI-83 Plus*) is called floor in SCHEME. And where LOGO uses FIRST and BUTFIRST to find the first element of a list and to remove the first element from a list, respectively, SCHEME uses car and cdr, abbreviations for "contents of address register" and "contents of decrement register", terms referring not to mathematical concepts, but to the original

---

[1] Generally, URLs that I find in print don't work by the time I read them and I am reluctant to present the URL here for fear of causing a change. Nonetheless, here it is: http://www.eecs.berkeley.edu/~bh/.

implementation of SCHEME's parent language LISP on the *IBM 704*.[2] Then too there is the archaism factor: LISP was developed on a machine with a more limited keyboard than we have today. Thus, one sees things like <= for ≤ and −> for →.

As is the case with LOGO, freeware versions of SCHEME can be found for the major platforms online. I use DRSCHEME, which is part of the PLT SCHEME package.[3]

One more thing must be said before presenting any programs. Each computer language is based on some *metaphor*. The *TI-83 Plus* has a *procedural* language: programs are step-by-step instructions on how to proceed. LOGO is largely *object-oriented*; it originated with a small device called a turtle that one gave instructions to. SCHEME is a *functional* language: every statement is an expression to be evaluated. The metaphor affects the style of one's programs. For example, in the SCHEME programs we will present below we will not use For loops as we did on the *TI-83 Plus*, but will use a recursive procedure simulating such.

I begin with a few simple programs written in SCHEME:

```
(define (square x)
   (* x x))
(define (int x)
   (inexact->exact (floor x)))
(define (number.of.digits.of n)
   (if (< n 10)
      1
      (+ 1 (number.of.digits.of (quotient n 10)))))
(define (adjoin.00s.to n p)
   (* n (expt 100 p)))
```

The first lines,

(define (*program.name variables*)),

are similar to LOGO's

TO *program.name :variables*.

They assign names to functions of the listed variables, which take the ensuing expressions as values.

Our first program is self-explanatory. The second overcomes a small glitch: The reader may recall that floor is SCHEME's name for the greatest integer function. It produces an integer, but as a real number, not an integer. Thus, if floor(x) is 6, it produces 6. or 6.0 or 6.00, depending on the format chosen for such. The SCHEME primitive inexact−>exact converts the real 6.0 to an integer

---

[2] Michael Eisenberg, William Clinger, Anne Hartheimer, and Harold Abelson, *Programming in MacScheme*, MIT Press, Cambridge (Mass.), 1990, p. 105. They add that "cdr is pronounced 'could-er', rhyming with 'good-er'".
[3] The URL is: http://plt-scheme.org/.

6 considered as an integer.[4] I should also quickly explain that expt is SCHEME's primitive command for exponentiation, its first argument being the base and its second the exponent of the operation.

Here is another program:

```
(define (length.100 n)
  (let* ([m (number.of.digits.of n)]
         [k (quotient m 2)]
         [r (- m (* 2 k))])
    (if (= r 0) k (+ k 1))))
```

(length.100 n) evaluates to the number of digits in the centesimal representation of n. The let command is the combination of an assignment of values to a list of variables and a declaration that the assignment is local, its scope being the command following the declaration list. Its overall structure is

```
(let ( [var₁ value₁]
       [var₂ value₂]
         ⋮
       [varₙ valueₙ])
    command)
```

There are two forms to the command: let and let*. The latter allows a variable var_i to be used in a value $value_j$ occurring later in the list of assignments, as we did with m and k.

As I said earlier, with exact integers, we don't need to use lists at all. However, the exact representation of integers loses leading 0's and it is convenient to store the number d whose square root we want as a list of centesimal digits:

```
(define (convert.to.base.100 n)
  (let ([k (length.100 n)])
    (build.conversion n k k)))

(define (build.conversion n k i)
  (if (= i 1)
      (list (remainder n 100))
      (cons (remainder (quotient n (expt 100 (- i 1))) 100)
            (build.conversion n k (- i 1)))))
```

This represents one way of implementing a For loop recursively. The program convert.to.base.100 takes an input n, finds its length k in base 100 and for $i = k$ to 1 creates the list by placing the i-th pair of base 10 digits of n in the front of a partial list of base 100 digits. The halting is handled by an if statement:

```
if (condition)
   (then-command)
   (else-command)),
```

---

[4] Note that the primitive is not written real—>integer because SCHEME also treats rational numbers exactly. Thus inexact—>exact 2.5 makes sense and evaluates to $\frac{5}{2}$.

which could be written on one line—the parentheses, not carriage returns determine the grouping. One more thing needs to be explained: the cons command is SCHEME's version of LOGO's FPUT. Its syntax is

(cons *value* *list*)

and it constructs[5] a new list by placing *value* first in a list and then following it with the members of *list* in their given order.

The square root program is now easy to write:

```
(define (square.root d p)
   (let* ([m (length.100 d)]
         [d.ext (adjoin.00s.to d p)]
         [d.list (convert.to.base.100 d.ext)]
         [c (+ m p)]
         [a (int (sqrt (car d.list)))]
         [diff (− (car d.list) (square a))])
      (fancy.print (square.root.aux a diff (cdr d.list) (− c 1)) d p)))

(define (square.root.aux curr.a curr.d curr.d.tail k)
   (if (= k 0)
      curr.a
      (let*([inter.a (∗ 10 curr.a)]
         ([next.d.tail (cdr curr.d.tail)]
         ([inter.d (+ (∗ 100 curr.d) (car curr.d.tail))]
         ([b (get.b inter.a inter.d)]
         ([next.a (+ inter.a b)]
         ([next.d (− inter.d (∗ b (+ (∗ 2 inter.a) b)))])
         ((square.root.aux next.a next.d next.d.tail (− k 1)))))))

(define (get.b a diff)
   (let ([b (quotient diff (∗ 2 a))])
      (if (= b 0)
         0
         (get.b.aux b (∗ 2 a) diff))))

(define (get.b.aux b c d)
   (if (>= (∗ b (+ c b)) d) (− b 1) b))

(define (fancy.print x d p)
   x)
```

These programs demand some explanation. We have a great deal more freedom in naming variables in SCHEME than in the programming language of the *TI-83 Plus*. However, I have decided to stick closely to the names used in the last chapter on the theory that this may be helpful. The variables m, c are the variables M, C of SQROOT. a is the value represented by the first occurrence of S in SQROOT and by *a* in our earlier discussion of the algorithm.

---

[5] cons is an abbreviation for "constructs". This is, of course, inconsistent with the naming practice used in choosing car and cdr.

d.ext ("ext" for "extended") is the numerical form of our former list ⌊D, which is here represented by d.list. And diff is the numerator $d - a^2$ that will be divided by $2a$ to get the next value of b (represented by the variable S in SQROOT). a and the succeeding b's are the successive digits of the square root. The program square.root then passes certain parameters to a recursive procedure square.root.aux that simulates our earlier For loop. We could leave it at that, but I chose to complicate matters by plugging the value obtained by calling square.root.aux into a program fancy.print. For the time being I have defined fancy.print to do nothing other than return the value of x it receives as input. One can later change fancy.print to re-insert the decimal point, print the result to a file, add formatting instructions for a table— whatever one pleases.

One part of the program strikes me as inelegant. That is the call to the function sqrt. A little more in keeping with the paper and pencil approach might be to replace the assignment by

[a (initial.root (car d.list))]

where we define

```
(define (initial.root n)
  (cond ((and (<= 1 n) (< n 4)) 1)
        ((and (<= 4 n) (< n 9)) 2)
        ((and (<= 9 n) (< n 16)) 3)
          ⋮
        ((and (<= 64 n) (< n 81)) 8)
        ((<= 81 n) 9)))
```

The command cond (for "conditional") allows one to define a function by several cases without having to nest if commands. Its structure is

```
(cond (case₁  command₁)
      (case₂  command₂)
         ⋮
      (caseₙ  commandₙ)
      (else  commandₙ₊₁))
```

The last line is optional. The cond command is evaluated as follows: First $case_1$ is tested. If it is true, $command_1$ is executed and the cond command is exited. Only if $case_1$ is false does one proceed to test $case_2$ and repeat the process. Thus, it is not necessary to guarantee that the cases are non-overlapping as I did and one can simplify the procedure initial.root by deleting the and's and the first conjuncts of each condition.

The program square.root.aux is the recursive part of the procedure and is best explained by looking back at FIGURE 1 of the preceding chapter. Once we used the first centesimal digit of d to get a we discarded it, and took the difference $d - a^2$ as a preliminary next value of d. a is fed to square.root.aux as curr.a, the current value of a; the difference diff is fed as curr.d, the current value of d; the remains of d.list after discarding the initial centesimal is fed as curr.d.tail, the current tail of d.list; and $c - 1$ starts a counter that will count down to 0. Back in FIGURE 1, we took the initial curr.a $= 4$, curr.d $=$ diff $= 18 - 4^2 = 2$, curr.d.tail $= \{18, 95, 74, 00, \ldots, 00\}$, and $c - 1 = 52$. Tacking the next centesimal 18 onto 2 amounted to multiplying 2 by 100 and adding 18, giving an intermediate value of d, inter.d. Multiplying a number by 100 multiplies its root by 10, whence inter.a. Also, we no longer need the 18, so we discard it from d.tail, forming the next value next.d.tail. We then have only to get the next b as in GETVARS, use it to construct the next value of a and the next value of d and feed next.a, next.d, and next.d.tail into square.root.aux, while lowering the counter by 1.

The program get.b differs somewhat from GETVARS, where b was generated by a testing loop. Here we have a quotient function giving us

$$\left[\frac{d - a^2}{2a}\right]$$

exactly, so we use that instead. Basically one then chooses b or $b-1$ or $\ldots$ as in the last line of code of get.b.aux. This does not work, however, if d is a perfect square. In that case, b can be 0 and without the first test one will choose $b - 1 = -1$. Thus, in finding $\sqrt{4}$, for example, the first step yields $a = 2$ as expected. But then get.b would produce -1 and in square.root.aux a becomes $2 - 1 = 1$. The iterated process goes on to generate

$$1.\underbrace{99\ldots9}_{p}$$

as the square root of 4. Using the outer if command, however, allows the simple correction of taking 0 instead of -1 in this case.

The program square.root works marvellously, producing the square root of a positive integer to p decimal places in short order. Redefining fancy.print to produce not the square root, but the square root together with LATEX formatting instructions, I was able to produce the following table of square roots of all whole numbers from 1 to 100, each to 50 decimals, in about a second[6] on my *iMac*. This is much better than the estimate of Project 11 of the last chapter for the time required on the *TI-83 Plus*.

TABLE 1: Square Roots to 50 Decimals

d	$\sqrt{d}$
1	1 . 00000000000000000000000000000000000000000000000000
2	1 . 41421356237309504880168872420969807856967187537694
3	1 . 73205080756887729352744634150587236694280525381038
4	2 . 00000000000000000000000000000000000000000000000000
5	2 . 23606797749978969640917366873127623544061835961152
6	2 . 44948974278317809819728407470589139196594748065667

---

[6] It scrolled too fast for me to time it with any degree of exactness.

d	$\sqrt{d}$
7	2 . 6457513110645905905016157536392604257102591830824 5
8	2 . 8284271247461900976033774484193961571393437507538 9
9	3 . 0000000000000000000000000000000000000000000000000 0
10	3 . 1622776601683793319988935444327185337195551393252 1
11	3 . 3166247903553998491149327366706866839270885455893 5
12	3 . 4641016151377545870548926830117447338856105076207 6
13	3 . 6055512754639892931192212674704959462512965738452 4
14	3 . 7416573867739413855837487323165493017560198077787 2
15	3 . 8729833462074168851792653997823996108329217052915 9
16	4 . 0000000000000000000000000000000000000000000000000 0
17	4 . 1231056256176605498214098555974077025147199225373 62
18	4 . 2426406871192851464050566172629094235709015626130 84
19	4 . 3588989435406735522369819838596156591370039252324 4
20	4 . 4721359549995793928183473374625524708812367192230 5
21	4 . 5825756949558400065880471937280084889844565767679 7
22	4 . 6904157598234295545656301135444662805882283534117 3
23	4 . 7958315233127195415974380641626939199967070419041 2
24	4 . 8989794855663561963945681494117827839318949613133 4
25	5 . 0000000000000000000000000000000000000000000000000 0
26	5 . 0990195135927848300282241090227819895637709460995 9
27	5 . 1961524227066318805823390245176171008284157614311 4
28	5 . 2915026221291811810032315072785208514205183661649 0
29	5 . 3851648071345040312507104915403295562951201616447 8
30	5 . 4772255750516611345696978280080213395274469499798 3
31	5 . 5677643628300219221194712989185495204763933775704 1
32	5 . 6568542494923801952067548968387923142786875015077 9
33	5 . 7445626465380286598506114682189293182202644579827 9
34	5 . 8309518948453004708741528775455830765213983348859 7
35	5 . 9160797830996160425673282915616170484155012307943 4
36	6 . 0000000000000000000000000000000000000000000000000 0
37	6 . 0827625302982196889996842452020670620849700947864 1
38	6 . 1644140029689764502501923814542442252356240234445 7
39	6 . 2449979983983982058468931209397944610729599779916 5
40	6 . 3245553203367586663977870888654370674391102786504 3
41	6 . 4031242374328486864882176746218132645204201326210 1
42	6 . 4807406984078602309659674360879966577052043070583 4
43	6 . 5574385243020006523441099976360016279269663198837 8
44	6 . 6332495807107996982298654733413733678541770911787 0
45	6 . 7082039324993690892275210061938287063218550788345 7
46	6 . 7823299831252681390645563266259691051957483239232 8
47	6 . 8556546004010441249358714490848489604606434610013 2

d	$\sqrt{d}$
48	6 . 9282032302755091741097853660234894677712210 1524152
49	7 . 0000000000000000000000000000000000000000000 0000000
50	7 . 0710678118654752440084436210484903928483593 7688474
51	7 . 1414284285428499799939981136726527876617115 990273
52	7 . 2111025550927978586238442534940991892502593 14769049
53	7 . 2801098892805182710973024915270327937776696 8257647
54	7 . 3484692283495342945918522241176741758978424 4197001
55	7 . 4161984870956629487113974408007130609799043 1909750
56	7 . 4833147735478827711674974646330986035120396 1555745
57	7 . 5498344352707496972366848069461170582221947 0462338
58	7 . 6157731058639082856614110271583230053607055 9254659
59	7 . 6811457478686081757696870217313724730624510 7761488
60	7 . 7459666924148337703585307995647992216658434 1058318
61	7 . 8102496759066543941297227357591014135683051 3664856
62	7 . 8740078740118110196850344488120078636810861 2202085
63	7 . 9372539331937717715048472609177812771307775 4924735
64	8 . 0000000000000000000000000000000000000000000 0000000
65	8 . 0622577482985496523666132303037711311343963 0560857
66	8 . 1240384046359603604598835682660403485042040 8672531
67	8 . 1853527718724499699537037247339294588804868 1549803
68	8 . 2462112512353210996428197119481540502943984 5074724
69	8 . 3066238629180748525842627449074920102322142 4895565
70	8 . 3666002653407554797817202578518748939281536 9298672
71	8 . 4261497731763586306341399062027360316080024 0156075
72	8 . 4852813742385702928101323452581884714180312 5226168
73	8 . 5440037453175311678716483262397064345944553 2953328
74	8 . 6023252670426267717294735350497136320275355 5729073
75	8 . 6602540378443864676372317075293618347140262 6905190
76	8 . 7177978870813471044739639677192313182740078 5046488
77	8 . 7749643873921220604063883074163095608758768 2755450
78	8 . 8317608663278468547640427269592539641746394 8093141
79	8 . 8881944173155888500914416754087278170764506 0372952
80	8 . 9442719099991587856366946749251049417624734 3844610
81	9 . 0000000000000000000000000000000000000000000 0000000
82	9 . 0553851381374166265738081669840664130521244 6409694
83	9 . 1104335791442988819456261046886691900991391 6826495
84	9 . 1651513899116800131760943874560169779689131 5353594
85	9 . 2195444572928873100022742817627931572468050 4872246
86	9 . 2736184954957037525164160739901746262634689 1207629
87	9 . 3273790530888150455544755423205569832762406 9419165
88	9 . 3808315196468591091312602270889325611764567 0682347

d	$\sqrt{d}$
89	9 . 43398113205660381132066037762264071698362263341512
90	9 . 48683298050513799599668063329815560115866541797565
91	9 . 53939201416945649152621586023226540254623425250545
92	9 . 59166304662543908319487612832538783999341408380825
93	9 . 64365076099295499576003104743266318390690369306325
94	9 . 69535971483265802814888115084531339365215098795467
95	9 . 74679434480896390683841319989960029925258390033749
96	9 . 79795897113271239278913629882356556786378992262668
97	9 . 84885780179610472174621141491762448169613628744276
98	9 . 89949493661166534161182106946788654998770312763863
99	9 . 94987437106619954734479821001206005178126563676806
100	10 . 00000000000000000000000000000000000000000000000000

Readers of the preceding chapter are, I am sure, anxious to know the value of the square root of 189574 to 50 decimal places and may be disappointed to see that it does not appear in this table. Well, take heart, I haven't forgotten this number. Its square root to 50 decimals is

$$435.40096462915651193202466530773242035061121435557669$$

1 Project. Is this better or worse than the estimate (5) of the last chapter? Carry out the division of (5) as far as 50 decimal places and compare with the above result.

For the truly adventurous I propose the following projects.

2 Project. Is SCHEME on a personal computer so fast that we can use an inefficient program and not notice the difference? Program the algorithm of Project 12 of Chapter 8 in SCHEME, this time calculating the full table of 100 square roots to make the comparison.

3 Project. I quote Augustus de Morgan (1806 - 1871):

Another instance of computation carried to a paradoxical length, in order to illustrate a method, is the solution of $x^3 - 2x = 5$, the example given of Newton's method, on which all improvements have been tested. In 1831, Fourier's posthumous work on equations showed 33 figures of solution, got with enormous labour. Thinking this a good opportunity to illustrate the superiority of the method of W.G. Horner, not yet known in France, and not much known in England, I proposed to one of my classes, in 1841, to beat Fourier on this point, as a Christmas exercise. I received several answers, agreeing with each other, to 50 places of decimals. In 1848, I repeated the proposal, requesting that 50 places might be exceeded. I obtained answers of 75, 65, 63, 58, 57, and

52 places. But one answer, by Mr. W. Harris Johnston, of Dundalk, and of the Excise Office, went to 101 decimal places.[7]

Repeat or exceed Mr. Johnston's accomplishment.

The second computational task I have for this chapter concerns the Pell equation. Some of the elementary school teachers in my class found the algebra behind Bhāskara's algorithm, as presented back in Chapter 4, a bit daunting, and, indeed, it is a little unmotivated. The presentation in terms of continued fractions is a bit easier to follow, but the fractions themselves are a bit mysterious and my discussion was a taste, not a general treatment. The Euclidean Algorithm provides a more motivated, if less efficient, algorithm than Bhāskara's, and generates the continued fraction expansion of $\sqrt{d}$ along the way. This was hinted at at the end of Chapter 5 when I pointed out that the application of the Euclidean Algorithm to the golden ratio ϕ and 1 never terminates, whence we concluded the irrationality of ϕ. This particular application was worked out more explicitly in Chapter 6 on pages 97 - 98. And we discussed it briefly regarding $\sqrt{2}$ in particular and $\sqrt{d}$ in general on pages 126 - 127 of Chapter 8. It is time we looked into this in detail, and wrote a program to generate the necessary details.

In theory, we could program the *TI-83 Plus* to solve the Pell equation via the Euclidean Algorithm. I haven't done this, not because I think it will be disappointingly slow, like SQROOT was, but because I like the ability to generate formatted files on the computer and have thus once again written the programs in SCHEME.

The procedure, as the reader may recall, goes as follows. To solve

$$X^2 - dY^2 = 1,$$

for some nonsquare number d, one applies the Euclidean Algorithm to $\sqrt{d}, 1$, first applying the Division Algorithm to find

$$\sqrt{d} = q_0 \cdot 1 + r_0, \quad 0 < r_0 < 1, \quad q_0 \text{ a whole number.}$$

One starts off easily enough: $q_0 = [\sqrt{d}]$ and $r_0 = \sqrt{d} - q_0$. One then divides 1 by r_0:

$$1 = q_1 \cdot r_0 + r_1, \quad 0 < r_1 < r_0, \quad q_1 \text{ a whole number,}$$

and thereafter divides r_n by r_{n+1}:

$$r_n = q_{n+2} \cdot r_{n+1} + r_{n+2}, \quad 0 < r_{n+2} < r_{n+1}, \quad q_{n+2} \text{ a whole number.}$$

Each q_n is a positive integer and each r_n can be written in the form $x_n + y_n\sqrt{d}$. Thus we can write

$$\sqrt{d} = q_0(1 + 0\sqrt{d}) + (-q_0 + \sqrt{d}), \quad 0 < -q_0 + \sqrt{d} < 1.$$

---

[7] Augustus de Morgan, *A Budget of Paradoxes*, 2 vols., 2nd ed., Open Court Publishing Company, Chicago and London, 1915, pp. 66 - 67.

And, setting $x_{-1} = 1, y_{-1} = 0, x_0 = -q_0, y_0 = 1$, we have

$$\sqrt{d} = q_0(x_{-1} + y_{-1}\sqrt{d}) + (x_0 + y_0\sqrt{d}),$$

and the iteration reads

$$x_{-1} + y_{-1}\sqrt{d} = q_1(x_0 + y_0\sqrt{d}) + (x_1 + y_1\sqrt{d})$$
$$x_0 + y_0\sqrt{d} = q_2(x_1 + y_1\sqrt{d}) + (x_2 + y_2\sqrt{d})$$
$$\vdots$$

with all the necessary inequalities suppressed for the sake of readability. Because d is assumed not to be a square, $\sqrt{d}$ is irrational and the algorithm goes on forever. However—and we haven't proven this anywhere in this book—it will eventually generate a remainder $x_n + y_n\sqrt{d}$ for which

$$x_n^2 - dy_n^2 = 1.$$

Generally, one stops applying the algorithm when this happens.

One might not want to stop immediately upon solving the Pell equation. Note that from

$$\sqrt{d} = q_0 + r_0$$

one has

$$\sqrt{d} = q_0 + \frac{1}{\dfrac{1}{r_0}}, \tag{1}$$

while

$$1 = q_1 r_0 + r_1$$

yields

$$\frac{1}{r_0} = q_1 + \frac{r_1}{r_0}. \tag{2}$$

Combining (1) and (2) we have

$$\sqrt{d} = q_0 + \frac{1}{q_1 + \dfrac{r_1}{r_0}}. \tag{3}$$

Similarly, from

$$r_n = q_{n+2} \cdot r_{n+1} + r_{n+2}$$

we conclude

$$\frac{r_n}{r_{n+1}} = q_{n+2} + \frac{r_{n+2}}{r_{n+1}}. \tag{4}$$

Successively mixing (3) and (4), we have

$$\sqrt{d} = q_0 + \cfrac{1}{q_1 + \cfrac{1}{q_2 + \cfrac{r_2}{r_1}}}$$

$$= q_0 + \cfrac{1}{q_1 + \cfrac{1}{q_2 + \cfrac{1}{q_3 + \cfrac{r_3}{r_2}}}}.$$

etc. The sequence $q_0, q_1, q_2, \ldots$ of quotients is thus just the sequence of numbers occurring in the simple continued fraction expansion of $\sqrt{d}$.

Thus, there are two halting strategies we could apply here. We could continue application of the Euclidean Algorithm until a solution to the Pell equation is found, or we could specify in advance some number m of steps to apply so as to determine the first m numbers in the simple continued fraction expansion of $\sqrt{d}$. It turns out that the first strategy suffices: The sequence of numbers occurring in the continued fraction expansion is ultimately periodic and is determined by $q_0, q_1, \ldots, q_{n-1}$, where $x_n^2 - dy_n^2 = \pm 1$. I illustrate this with the following application of the algorithm for $d = 7$:

$$\sqrt{7} = 2 \cdot 1 + (-2 + \sqrt{7})$$
$$1 = 1(-2 + \sqrt{7}) + (3 - \sqrt{7})$$
$$-2 + \sqrt{7} = 1(3 - \sqrt{7}) + (-5 + 2\sqrt{7})$$
$$3 - \sqrt{7} = 1(-5 + 2\sqrt{7}) + (8 - 3\sqrt{7})$$
$$-5 + 2\sqrt{7} = 4(8 - 3\sqrt{7}) + (-37 + 14\sqrt{7})$$
$$8 - 3\sqrt{7} = 1(-37 + 14\sqrt{7}) + (45 - 17\sqrt{7})$$
$$-37 + 14\sqrt{7} = 1(45 - 17\sqrt{7}) + (-82 + 31\sqrt{7})$$
$$45 - 17\sqrt{7} = 1(-82 + 31\sqrt{7}) + (127 - 48\sqrt{7})$$
$$-82 + 31\sqrt{7} = 4(127 - 48\sqrt{7}) + (-590 + 223\sqrt{7})$$

$$\vdots$$

Continuing further, one finds the sequence of quotients to be

$$2, 1, 1, 1, 4, 1, 1, 1, 4, 1, 1, 1, 4, \ldots,$$

with the $1, 1, 1, 4$ repeating forever. That is, the sequence of quotients is

$$q_0, q_1, q_2, q_3, q_4 = 2q_0, q_1, q_2, q_3, q_4 = 2q_0, \ldots$$

Notice in this case that for $(x_4, y_4) = (8, -3)$, one has

$$x_n^2 - dy_n^2 = 8^2 - 7 \cdot 3^2 = 64 - 63 = 1.$$

In general, for any nonsquare d, once one finds n such that $x_n^2 - dy_n^2 = \pm 1$, the sequence $q_0, q_1, \ldots$ of quotients, i.e., the sequence of numbers in the continued fraction expansion of $\sqrt{d}$ is

$$q_0, q_1, q_2, \ldots, q_{n-1}, q_n = 2q_0, q_1, q_2, \ldots, q_{n-1}, q_n = 2q_0, \ldots$$

From a practical standpoint, this means that one does not have to continue application of the algorithm beyond the point where a solution to the Pell equation has been found, unless it is to see for oneself that this behaviour actually occurs. I have thus opted to stop the execution of the algorithm once a solution to the Pell equation has been found.

Writing the actual program is, except for one subtle point, fairly easy. Deciding among the possibilities is the hard part. Essentially, we want to go from an equation

$$r_n = q_{n+2} r_{n+1} + r_{n+2}$$

to

$$r_{n+1} = q_{n+3} r_{n+2} + r_{n+3},$$

where each r_m is of the form $x_m + y_m \sqrt{d}$. Essentially, this means we want to program the transition

$$q_{n+2}, r_{n+1}, r_{n+2} \mapsto q_{n+3}, r_{n+2}, r_{n+3}$$

or

$$q_{n+2}, x_{n+1}, y_{n+1}, x_{n+2}, y_{n+2} \mapsto q_{n+3}, x_{n+2}, y_{n+2}, x_{n+3}, y_{n+3}$$

and we could treat an input or output as one number and two lists of length 2, or as 5 numbers, or even as a single list of 5 numbers. I have decided to use this last alternative.

SCHEME provides us with the means of extracting the individual elements of a list. I am not sure how universal this is[8], but there are primitive SCHEME functions first, second, third, fourth, fifth singling out the first, second, third, fourth, and fifth elements of a list. If there weren't, or if one has an implementation lacking these, they can easily be defined as follows:

```
(define (first quintuple)
  (car quintuple))

(define (second quintuple)
  (car (cdr quintuple)))

(define (third quintuple)
  (car (cdr (cdr quintuple))))
```

---

[8] The PLT SCHEME implementation on my computer has these; but the book (Eisenberg, *et al.*, *op. cit.*) on MACSCHEME I learned the language from does not list these primitives. Indeed, it offers the following program for first:

```
(define (first car)).
```

```
(define (fourth quintuple)
  (car (cdr (cdr (cdr quintuple)))))
```

```
(define (fifth quintuple)
  (car (cdr (cdr (cdr (cdr quintuple)))))).
```

Other simple auxiliary functions we shall use are

```
(define (int x)
  (inexact->exact (floor x)))
```

```
(define (square x)
  (* x x))
```

```
(define (pell x y d)
  (- (square x) (* d (square y))))
```

Building quintuples is handled by a pair of programs. From the equation

$$\sqrt{d} = q_0(1 + 0\sqrt{d}) + (-q_0 + \sqrt{d})$$

we get

```
(define (first.quint d)
  (let ([q0 (int (sqrt d))])
    (list q0 1 0 (- q0) 1)))
```

And the transition from the quintuple $(q_{n+1}, x_n, y_n, x_{n+1}, y_{n+1})$ to the quintuple $(q_{n+2}, x_{n+1}, y_{n+1}, x_{n+2}, y_{n+2})$ based on

$$x_n + y_n\sqrt{d} = q_{n+2}(x_{n+1} + y_{n+1}\sqrt{d}) + (x_{n+2} + y_{n+2}\sqrt{d})$$

requires

$$q_{n+2} = \left\lfloor \frac{x_n + y_n\sqrt{d}}{x_{n+1} + y_{n+1}\sqrt{d}} \right\rfloor$$

$$x_{n+2} = x_n - q_{n+2}x_{n+1}$$

$$y_{n+2} = y_n - q_{n+2}y_{n+1}.$$

This translates to

```
(define (next.quint quint d)
  (let* ([x (second quint)]
         [y (third quint)]
         [z (fourth quint)]
         [w (fifth quint)]
         [new.q (next.q x y z w d)]
         [new.z (- x (* new.q z))]
         [new.w (- y (* new.q w))])
    (list new.q z w new.z new.w)))
```

The tricky part is defining the function next.q, which is to be the greatest integer in the ratio

$$\frac{x + y\sqrt{d}}{z + w\sqrt{d}}.$$

Now, neither the numerator nor the denominator is an integer, whence quotient does not apply. And, as we proceed further in the iteration, both numerator and denominator are very small and rounded off and the computer quickly starts giving incorrect values. To get around this, we must work solely with rational numbers, and in older implementations of SCHEME with integers. To this end, rationalise the denominator:

$$\frac{x + y\sqrt{d}}{z + w\sqrt{d}} = \frac{(x + y\sqrt{d})(z - w\sqrt{d})}{(z + w\sqrt{d})(z - w\sqrt{d})}$$

$$= \frac{(xz - dyw) + (yz - xw)\sqrt{d}}{z^2 - dw^2}.$$

Now, if

$$n = \left\lfloor \frac{x + y\sqrt{d}}{z + w\sqrt{d}} \right\rfloor,$$

then

$$n < \frac{a + b\sqrt{d}}{c} < n + 1, \tag{5}$$

where $a = xz - dyw$, $b = yz - xw$, and $c = z^2 - dw^2$. We find n by testing (5) for $k = 1, 2, \ldots$ until n is found. The program next.q sets things up and calls an iteration program next.q.aux to do the successive tests:

```
(define (next.q x y z w d)
  (let ([a (− (∗ x z) (∗ d (∗ y w)))]
        [b (− (∗ y z) (∗ x w))]
        [c (pell z w d)])
    (next.q.aux a b c d 1)))
```

Testing inequalities between expressions $p + q\sqrt{d}, r + s\sqrt{d}$, with p, q, r, s rational, is an easy matter:

$$p + q\sqrt{d} < r + s\sqrt{d} \quad \text{iff} \quad p - r < (s - q)\sqrt{d},$$

so it suffices to be able to test inequalities of the form $m < n\sqrt{d}$ with m, n rational. This is handled simply by considering the various cases:

$$m < n\sqrt{d} \quad \text{iff:} \quad m < 0 \ \& \ n < 0 \ \& \ m^2 > n^2 d, \ \text{or}$$
$$m < 0 \ \& \ 0 \le n, \ \text{or}$$
$$0 \le m \ \& \ 0 \le n \ \& \ m^2 < n^2 d.$$

Thus we have

```
(define (less.than p q r s d)
  (let* ([m (− p r)]
         [n (− s q)]
         [M (square m)]
         [N (* d (square n))])
    (cond ((and (< m 0) (< n 0) (< N M)) #t)
          ((and (< m 0) (<= 0 n) #t)
          ((and (<= 0 m) (<= 0 n) (< M N)) #t)
          (else #f))))).
```

The values #t and #f are the truth values *true* and *false*, respectively. Since we are producing truth values as output here, we could equally well replace cond by or and delete the final else command. Also, since we will not be interested in the cases where m or n is 0, the <= commands can be replaced by < commands. With all of this, next.q.aux is easily defined:

```
(define (next.q.aux a b c d k)
  (let ([A (/ a c)]
        [B (/ b c)])
    (if (and (less.than k 0 A B d) (less.than A B (+ k 1) 0 d))
        k
        (next.q.aux a b c d (+ k 1))))))).
```

If one has an older implementation of SCHEME which does not handle rational numbers exactly, one must rewrite (5) using only integers by multiplying by the absolute value of *c*. In this case, next.q.aux will read

```
(define (next.q.aux a b c d k)
  (let ([A (* a (abs c))]
        [B (* b (abs c))])
    (if (and (less.than (* (abs c) k) 0 A B d)
             (less.than A B (* (abs c) (+ k 1)) 0 d))
        k
        (next.q.aux a b c d (+ k 1))))))).
```

The rest is now fairly easy and depends only on what we want. Are we interested in displaying full information for a particular *d*, or just the fundamental solution to the Pell equation $X^2 - dY^2 = 1$? Do we want to make a table of such solutions to the Pell equation?

The full solution to an instance of the Pell equation should display, for each *n*, the numbers q_n, x_n, y_n and the pellian value $x_n^2 - dy_n^2$.

```
(define (full.pell.info d)
  (let* ([fq (first.quint d)]
         [x0 (fourth fq)]
         [y0 (fifth fq)])
    (display " − ")
    (display 1)
    (display " ")
```

```
(display (list 1 0))
(newline)
(display (first fq))
(display " ")
(display (pell x0 y0 d))
(display " ")
(display (list x0 y0))
(newline)
(full.pell.aux 1 0 x0 y0 d)))

(define (full.pell.aux x y z w d)
  (let ([p (pell z w d)])
    (if (= p 1)
        (display " ")
        (let* ([new.quint (next.quint (list 0 x y z w) d)]
               [new.z (fourth new.quint)]
               [new.w (fifth new.quint)])
          (display (first new.quint))
          (display " ")
          (display (pell new.z new.w d))
          (display " ")
          (display (list new.z new.w))
          (newline)
          (full.pell.aux z w new.z new.w d))))))).
```

Adding to the display commands of full.pell.info and full.pell.aux extra commands displaying LATEX formatting instructions, I generated the tables to follow. I have chosen the values $d = 13, 19$ (these values just as examples), 61, 92, 109, and 149 (historical values: 92 is, of course, Brahmagupta's example, the one he said could be solved in less than a year only by a mathematician; and the others were explicitly mentioned by Fermat as being particularly difficult[9]). There are other examples worth examining, in particular our old friend $d = 189574$ and the pair $d = 1620, 1621$. I forego the pleasure of presenting their tables here because the sizes of the solutions would require me to reformat them so as to fit the width of the page. The value $d = 189574$ is interesting because the values $-1, \pm 2, \pm 4$ never occur as values in the $x^2 - dy^2$ column and thus the obvious simplifications in computation cannot be made when performing the calculations by hand. And the pair $1620, 1621$ is interesting because $d = 1620$

---

[9] Whitford (*The Pell Equation*, Merchant books, 2008, p. 59) quotes a letter from Euler to Goldbach: "Problems of this kind have been discussed between Wallis and Fermat... The most difficult example was to find numbers which put in place of y would make $109y^2 + 1$ a perfect square." Another difficult example cited by Fermat is $d = 433$. Indeed, it was the most difficult case solved by Fermat (Whitford: p. 52), and, along with $d = 313$, one of the most difficult solved by Wallis and Brouncker (Whitford: pp. 57, 59). "Difficult" here means merely that the solution is very large.

$d = 61$

q_n	$x_n^2 - dy_n^2$	x	y
—	1	1	0
7	−12	−7	1
1	3	8	−1
4	−4	−39	5
3	9	125	−16
1	−5	−164	21
2	5	453	−58
2	−9	−1070	137
1	4	1523	−195
3	−3	−5639	722
4	12	24079	−3083
1	−1	−29718	3805
14	12	440131	−56353
1	−3	−469849	60158
4	4	2319527	−296985
3	−9	−7428430	951113
1	5	9747957	−1248098
2	−5	−26924344	3447309
2	9	63596645	−8142716
1	−4	−90520989	11590025
3	3	335159612	−42912791
4	−12	−1431159437	183241189
1	1	1766319049	−226153980

$d = 13$

q_n	$x_n^2 - dy_n^2$	x	y
—	1	1	0
3	−4	−3	1
1	3	4	−1
1	−3	−7	2
1	4	11	−3
1	−1	−18	5
6	4	119	−33
1	−3	−137	38
1	3	256	−71
1	−4	−393	109
1	1	649	−180

$d = 19$

q_n	$x_n^2 - dy_n^2$	x	y
—	1	1	0
4	−3	−4	1
2	5	9	−2
1	−2	−13	3
3	5	48	−11
1	−3	−61	14
2	1	170	−39

TABLE 2. Solutions to Selected Pell Equations

is solved ($x = 161, y = 4$) in only 2 steps, while $d = 1621$ requires over 150 steps and results in numbers of 70+ digits.

Of these tables, those for $d = 19$ and $d = 92$ are particularly interesting. The former has no −1 in its pellian column, but does have a −2, while the latter has neither. But it does have a 4. In fact, all the examples given here have a 4 or a −4. The table, not exhibited here, for $d = 189574$ has none of these simplifying values.

$$d = 92$$

q_n	$x_n^2 - dy_n^2$	x	y
–	1	1	0
9	−11	−9	1
1	8	10	−1
1	−7	−19	2
2	4	48	−5
4	−7	−211	22
2	8	470	−49
1	−11	−681	71
1	1	1151	−120

TABLE 3. Solution to the Pell Equation for $d = 92$

As one looks over these tables, one finds the x- and y-columns not so interesting. One might notice that they alternate from step to step from x being positive and y being negative to x being negative and y being positive, and *vice versa*, but that is as interesting as it gets. The q- and pellian columns are more interesting. I have stopped the tables as soon as a 1 re-appeared in the pellian column, i.e., as soon as the fundamental solution $x_n^2 - dy_n^2 = 1$ was found, so the tables don't exhibit it, but these sequences are periodic. In fact, if $X^2 - dY^2 = -1$ has no solution, the full period $q_1, \ldots, q_n$ of the q's is not exhibited as $q_n = 2q_0$ would first occur in the next line after extending the table. However, what does appear is remarkable. The sequence $q_1, \ldots, q_{n-1}$, i.e., that part of the repeating numbers other than the last, is a palindrome, reading the same backwards:

$$(q_1, \ldots, q_{n-1}) = (q_{n-1}, \ldots, q_1).$$

And, writing $p_0, p_1, \ldots$ for the elements of the pellian column, one also gets a palindromic period, and the successive elements of the sequence alternate in sign. If you look at enough examples, you should see more.

$$d = 109$$

q_n	$x_n^2 - dy_n^2$	x	y
–	1	1	0
10	−9	−10	1
2	5	21	−2
3	−12	−73	7
1	7	94	−9
2	−4	−261	25

q_n	$x_n^2 - dy_n^2$	x	y
4	15	1138	−109
1	−3	−1399	134
6	3	9532	−913
6	−15	−58591	5612
1	4	68123	−6525
4	−7	−331083	31712
2	12	730289	−69949
1	−5	−1061372	101661
3	9	3914405	−374932
2	−1	−8890182	851525
20	9	181718045	−17405432
2	−5	−372326272	35662389
3	12	1298696861	−124392599
1	−7	−1671023133	160054988
2	4	4640743127	−444502575
4	−15	−20233995641	1938065288
1	3	24874738768	−2382567863
6	−3	−169482428249	16233472466
6	15	1041769308262	−99783402659
1	−4	−1211251736511	116016875125
4	7	5886776254306	−563850903159
2	−12	−12984804245123	1243718681443
1	5	18871580499429	−1807569584602
3	−9	−69599545743410	6666427435249
2	1	158070671986249	−15140424455100

TABLE 4: Solution to the Pell Equation for $d = 109$

$d = 149$

q_n	$x_n^2 - dy_n^2$	x	y
−	1	1	0
12	−5	−12	1
4	17	49	−4
1	−4	−61	5
5	7	354	−29
3	−7	−1123	92
3	4	3723	−305
5	−17	−19738	1617

q_n	$x_n^2 - dy_n^2$	x	y
1	5	23461	-1922
4	-1	-113582	9305
24	5	2749429	-225242
4	-17	-11111298	910273
1	4	13860727	-1135515
5	-7	-80414933	6587848
3	7	255105526	-20899059
3	-4	-845731511	69285025
5	17	4483763081	-367324184
1	-5	-5329494592	436609209
4	1	25801741449	-2113761020

TABLE 5: Solution to the Pell Equation for $d = 149$

4 Project. Insert the command

(display (list a b c))

into the program next.q just before the call to next.q.aux and generate some tables. What do you notice? (For the more advanced mathematician:) Can you prove that this always happens? My original suite of programs used the following program in place of the current pair less.than and next.q.aux:

```
(define (next.q.aux a b c d k)
  (let ([left (square (- (* k c) a))]
        [right (square (-(* (+ k 1) c) a))]
        [center (* d (square b))])
    (if (and (< left center)(< center right))
      k
      (next.q.aux a b c d (+ k 1)))))
```

Can you prove that this will always work? Can you give a proof that would be appropriate at the level this book is trying to be written at?

If we don't want the full information, but just the solution to the Pell equation, we can rewrite our last two programs to output values rather than display lots of information:

```
(define (pell.solver d)
  (let* ([fq (first.quint d)]
         [x0 (fourth fq)]
         [y0 (fifth fq)])
    (pell.solver.aux 1 0 x0 y0 d)))

(define (pell.solver.aux x y z w d)
  (let ([p (pell z w d)])
```

```
(if (= p 1)
  (list z w)
  (let* ([new.quint (next.quint (list 0 x y z w) d)]
     [new.z (fourth new.quint)]
     [new.w (fifth new.quint)])
    (pell.solver.aux z w new.z new.w d))))).
```

The program pell.solver will do the trick. It is not very efficient, but modern computers are so fast that it won't matter unless one is making a large table of solutions and the extra time accumulates. One would certainly notice it on the calculator or in performing calculations by hand. Consider, for example, the table for $d = 61$. The 12th entry yields

$$29718^2 - 61 \cdot 3805^2 = -1.$$

Instead of proceeding 12 more steps, one would simply apply Brahmagupta's multiplication rule to $29718 + 3805\sqrt{61}$:

$$(29718 + 3805\sqrt{61})^2 = (29718^2 + 61 \cdot 3805^2) + (2 \cdot 29718 \cdot 3805)\sqrt{61}$$
$$= 1766319049 + 226153980\sqrt{61}$$

and thus obtain the solution

$$1766319049^2 - 61 \cdot 226153980^2 = 1$$

in one step. Another simplification can be made for $d = 54$, where -2 occurs as the pellian value in the fourth line:

$$22^2 - 54 \cdot 3^2 = -2.$$

The program requires 3 more steps to obtain

$$485^2 - 54 \cdot 66^2 = 1,$$

but we can do it in one step if we again follow Brahmagupta:

$$(22 + 3\sqrt{54})^2 = (22^2 + 9 \cdot 54) + (2 \cdot 22 \cdot 3)\sqrt{54}$$
$$= 970 + 132\sqrt{54}.$$

Division by 2 yields $485 + 66\sqrt{54}$, i.e., the solution $(485, 66)$.

A far more common occurrence than this is a ± 4 in the pellian column. This case is a bit more complicated, but handling it often eliminates over half the steps in the computation. Going back to $d = 61$ as an example, we might note from the 4th row of the table that

$$39^2 - 61 \cdot 5^2 = -4.$$

Applying Brahmagupta's multiplication, we get

$$(39 + 5\sqrt{61})^2 = (39^2 + 61 \cdot 5^2) + (2 \cdot 39 \cdot 5)\sqrt{61}$$

$$= 3046 + 390\sqrt{61}$$

with even values of x, y. Dividing by 2 yields $1523 + 195\sqrt{61}$ with

$$1523^2 - 61 \cdot 195^2 = 4.$$

Multiplying by $39 + 5\sqrt{61}$ again yields

$$(39 + 5\sqrt{61})(1523 + 195\sqrt{61}) = 118872 + 15220\sqrt{61}.$$

Dividing by 2 yields $59436 + 7610\sqrt{61}$, which is again even, whence one divides once more by 2 to get $29718 + 3805\sqrt{61}$. Now this has a pellian value of -1 and we have already seen that squaring it yields the solution. Note that this essentially added 3 more cycles to the 4 yielding the pair $(39, 5)$, thus 7 in all as opposed to the full iteration of 23 cycles. More than $2/3$ of the work has been eliminated.

Of course, saving steps in calculation requires adding steps in programs, in this case in pell.solver.aux:

```
(define (pell.solver.aux x y z w d)
  (let ([p (pell z w d)])
    (cond ((= p 1) (list z w))
      ((= p (- 1)) (pell.aux.1 z w d))
      ((or (= p 2) (= p (- 2))) (pell.aux.2 z w d))
      ((or (= p 4) (= p (- 4))) (pell.aux.4 z w d))
      (else (let* ([new.quint (next.quint (list 0 x y z w) d)]
            [new.z (fourth new.quint)]
            [new.w (fifth new.quint)])
        (pell.solver.aux z w new.z new.w d)))))))
```

The programs pell.aux.1 and pell.aux.2 are simple enough:

```
(define (pell.aux.1 x y d)
  (let ([z (+ (square x) (* d (square y)))]
    [w (* 2 (* x y))])
  (list z w)))
```

```
(define (pell.aux.2 x y d)
  (let* ([z1 (+ (square x) (* d (square y)))]
    [z (quotient z1 2)]
    [w (* x y)])
  (list z w)))
```

The program pell.aux.4 is a little more complicated, but not too much so. It turns out one does not have to apply Brahmagupta's multiplication trick repeatedly. Success comes in 1, 2 or 3 stages.

Let us examine the situation more closely. If

$$x^2 - dy^2 = 4\varepsilon, \quad \text{with } \varepsilon = \pm 1,$$

we multiply $x + y\sqrt{d}$ by itself to get

$$(x + y\sqrt{d})^2 = (x^2 + dy^2) + (2xy)\sqrt{d}.$$

Now

$$x^2 + dy^2 = x^2 - dy^2 + 2dy^2 = 4\varepsilon + 2dy^2$$

is even, as is $2xy$. We can thus divide by 2 to obtain

$$(2\varepsilon + dy^2)^2 - d(xy)^2 = \left(\frac{x^2 + dy^2}{2}\right)^2 - d\left(\frac{2xy}{2}\right)^2$$

$$= \frac{1}{4}\left((x^2 + dy^2)^2 - d(2xy)^2\right)$$

$$= \frac{1}{4}(4\varepsilon)^2 = 4\varepsilon^2 = 4.$$

If x or y is even, then $dy^2 = x^2 - 4\varepsilon$ and xy are also even and we can divide by 2 again to obtain integers

$$\varepsilon + \frac{dy^2}{2}, \frac{xy}{2} \quad \text{satisfying} \quad \left(\varepsilon + \frac{dy^2}{2}\right)^2 - d\left(\frac{xy}{2}\right)^2 = 1.$$

If x, y are both odd, we multiply again by $x + y\sqrt{d}$:

$$(x + y\sqrt{d})\left((2\varepsilon + dy^2) + xy\sqrt{d}\right)$$

$$= \left((2\varepsilon x + xdy^2 + xy^2 d) + (x^2 y + 2\varepsilon y + dy^3)\sqrt{d}\right)$$

$$= \left(2\varepsilon x + 2x(x^2 - 4\varepsilon)\right) + y\left(x^2 + 2\varepsilon + x^2 - 4\varepsilon\right)\sqrt{d}$$

$$= (2x^3 - 6\varepsilon x) + y(2x^2 - 2\varepsilon)\sqrt{d}$$

$$= 2x(x^2 - 3\varepsilon) + 2y(x^2 - \varepsilon)\sqrt{d}.$$

Because x is odd, $x^2 - 3\varepsilon = x^2 \mp 3$ and $x^2 - \varepsilon = x^2 \mp 1$ are even and we can divide $2x(x^2 - 3\varepsilon)$ and $2y(x^2 - \varepsilon)$ by 4 to obtain integers satisfying

$$\left(\frac{2x(x^2 - 3\varepsilon)}{4}\right)^2 - d\left(\frac{2y(x^2 - \varepsilon)}{4}\right)^2 = \frac{4\varepsilon \cdot 4}{4^2} = \varepsilon.$$

If $\varepsilon = 1$, we have the solution we are looking for, and, if $\varepsilon = -1$, we have only to apply pell.aux.1 to obtain it. We embody this information in a program as follows:

```
(define (pell.aux.4 x y d)
   (let* ([X1 (+ (square x) (* d (square y)))]
          [x1 (/ X1 2)]
          [y1 (* x y)])
      (if (or (even x) (even y))
          (list (/ x1 2) (/ y1 2))
          (let* ([X2 (+ (* x x1) (* d (* y y1)))]
                 [Y2 (+ (* x y1) (* x1 y))]
                 [x2 (/ X2 4)]
```

```
        [y2 (/ Y2 4)]
        [p (pell x2 y2 d)])
     (if (= p 1)
        (list x2 y2)
        (pell.aux.1 x2 y2 d))))))))
(define (even x)
  (= (remainder x 2) 0))
```

With these programs we can now fairly quickly calculate the fundamental solutions to the Pell equation for various values of d. I cite, for example the values $d = 313$ and $d = 433$ considered by Wallis and Brouncker, along with the value $d = 1621$ mentioned above:

$d = 313$
$x = 32188120829134849$
$y = 1819380158564160$

$d = 433$
$x = 104564907854286695713$
$y = 5025068784834899736$

$d = 1621$
$x = 6298101812493732343034974500091457815529942308667051412$
$\qquad$ 857352310169665125001
$y = 1564293243699791121284455833450983386275520438748241083$
$\qquad$ 9917792442751050500.

Applying pell.solver successively, we can quickly generate a table of solutions for however many values of d we choose. The following table, for $d \leq 100$ seemed to take less than a second to generate. I could have generated a much larger table in real time, but when the solutions start getting too large (e.g., $d = 1621$ as above), one has to take greater care in formatting. The present table does not offer anything too dramatic, but it does reveal the erratic nature of the solutions. The x and y values jump around, with relatively small pairs placed next to inexplicably large ones.

TABLE 6: Solutions to the Pell Equation

d	x	y	d	x	y
1	–	–	51	50	7
2	3	2	52	649	90
3	2	1	53	66249	9100
4	–	–	54	485	66
5	9	4	55	89	12
6	5	2	56	15	2
7	8	3	57	151	20
8	3	1	58	19603	2574
9	–	–	59	530	69
10	19	6	60	31	4

d	x	y	d	x	y
11	10	3	61	1766319049	226153980
12	7	2	62	63	8
13	649	180	63	8	1
14	15	4	64	–	–
15	4	1	65	129	16
16	–	–	66	65	8
17	33	8	67	48842	5967
18	17	4	68	33	4
19	170	39	69	7775	936
20	9	2	70	251	30
21	55	12	71	3480	413
22	197	42	72	17	2
23	24	5	73	2281249	267000
24	5	1	74	3699	430
25	–	–	75	26	3
26	51	10	76	57799	6630
27	26	5	77	351	40
28	127	24	78	53	6
29	9801	1820	79	80	9
30	11	2	80	9	1
31	1520	273	81	–	–
32	17	3	82	163	18
33	23	4	83	82	9
34	35	6	84	55	6
35	6	1	85	285769	30996
36	–	–	86	10405	1122
37	73	12	87	28	3
38	37	6	88	197	21
39	25	4	89	500001	53000
40	19	3	90	19	2
41	2049	320	91	1574	165
42	13	2	92	1151	120
43	3482	531	93	12151	1260
44	199	30	94	2143295	221064
45	161	24	95	39	4
46	24335	3588	96	49	5
47	48	7	97	62809633	6377352
48	7	1	98	99	10
49	–	–	99	10	1
50	99	14	100	–	–

Such tables of solutions to the Pell equation started appearing in Europe from the beginning of modern European interest in the equation.[10] Fermat himself found solutions for all nonsquare $d \leq 150$ and Euler included a table for $d \leq 100$ in his *Elements of Algebra* in 1770. Adrien Marie Legendre (1752 - 1833) presented a table for $d \leq 1003$ in his *Essai sur la théorie des nombres* in 1798, but it contained 38 errors. In 1817 Carl Ferdinand Degen (1766 - 1825) published a table of solutions for $d \leq 1000$, and this was complemented by the British Association in 1893 when it published a continuation to cover the values of d between 1001 and 1500. Finally, in 1911, in his book on the Pell equation, Whitford extended this further to include all values of d between 1501 and 1700. Presumably it was in perusing these tables that he noticed the remarkably different solutions to the equations for $d = 1620$ and $d = 1621$.

Fermat's letters to Bernard Frénicle de Bessy and Kenelm Digby were written in 1657. From then until 1911 is just over two and a half centuries in which to solve 1659 instances of the Pell equation[11]. With our modern high-speed computers, this can be done in considerably less time. Using the rather imprecise method of counting "one thousand and one, one thousand and two,...", I estimated that it took my computer 5 seconds to generate a formatted table (albeit one for a much wider page) of the solutions for all $d \leq 1700$. [12]

Having just completed duplicating the work of $2\frac{1}{2}$ centuries in 5 seconds, I am understandably feeling a bit tired, so I shall bring this chapter to a rapid close. I just want to add that, while the computer does speed things up, the great disparity in size between some values of d and its solution x, y means that the amount of computation involved will frequently be horrendous for multidigit numbers. Not knowing in advance how quickly the computer would solve the first 1700 cases of the Pell equation, I made no allowance for such large values of d in writing my programs. If one wants to use the program to solve the Pell equation for d with more digits than real numbers are allowed to have in SCHEME, one must replace the program first.quint by the following:

```
(define (first.quint d)
  (let ([q0 (inexact->exact (square.root d 0))])
    (list q0 1 0 (- q0) 1)),
```

and load the full suite of programs accompanying square.root. I leave this task and accompanying experimentation to the reader.

---

[10] Whitford, *op.cit.*, offers a brief outline of the history of such tables. The reference to Fermat is on p. 57, and I refer the reader to pp. 95 - 98 for the rest.

[11] $1659 = 1700 - [\sqrt{1700}] =$ the number of nonsquares ≤ 1700.

[12] And here I must confess I took the lazy approach of simply writing a program to call pell.solver 1700 times and format the answers. This is not the most efficient procedure. For each d, for example, the square root of d is calculated twice— once in the program first.quint and again in the iteration where it is calculated with an eye to determining whether or not d is a square.

10

Tables

Tables have been an aid to computation throughout all of recorded mathematical history. Egyptian papyri feature tables, most famously tables for doubling unit fractions, i.e., for representing fractions $2/n$ as sums of fractions of the form $1/m$, e.g., $2/5 = 1/3 + 1/15$. The more numerous Babylonian finds include a greater variety of tables: multiplication tables, reciprocals, squares and square roots, and the famous tablet Plimpton 322 offers a table of Pythagorean triples.

The purpose of tables such as these is fairly obvious: Computation can be a laborious procedure and if one carries out the work in advance and stores it in a table, one will not have to repeat a given calculation over and over. Astronomical tables, collecting observations, were also made. Such tables serve a different purpose, as the results cannot truly be calculated until after a theory is in place—and then, of course, one has two types of tables: observed and calculated values. The theory might arise from an examination of the observed data or it might come from some model proposed to explain the phenomena, the model providing some mathematical formula allowing one to calculate predicted values of the data.

Trigonometric tables, the origins of which are closely tied to Astronomy, have a long and proud history. And they have names associated with them.

Hipparchus calculated a table of chords for intervals of $7\frac{1}{2}$ degrees. It is reported he started with the directly calculable chords for $60°$ and $90°$, and used the half-angle formula

$$\text{chord}\left(\frac{\alpha}{2}\right) = \sqrt{\frac{1}{2}\left(d^2 - d\sqrt{d^2 - s^2}\right)}$$

for d the diameter of a given circle and $s = \text{chord}\,\alpha$. This gave him the chords of

$$90, 60, 45, 30, 22\tfrac{1}{2}, 15, 7\tfrac{1}{2}$$

degrees. Using

$$\text{chord}(180 - \alpha) = \sqrt{d^2 - s^2},$$

he could also calculate the chords of

$$172\tfrac{1}{2},\ 165,\ 157\tfrac{1}{2},\ 150,\ 135,\ 120$$

degrees. The rest of the values he obtained by *linear interpolation*.

Hipparchus based all of this on a circle of circumference $360°$, each degree consisting of 60 minutes. This gave a circle of circumference $360 \cdot 60$ minutes and he chose a radius

$$\frac{360 \cdot 60}{2\pi} \approx 3438'.$$

Note that the famous Archimedean value of π, namely $\tfrac{22}{7}$, would yield the estimate

$$\frac{360 \cdot 60}{2 \cdot \frac{22}{7}} \approx 3436.\overline{36} \approx 3436.$$

Hipparchus thus used a better estimate for π than this. It is believed he used that of Apollonius (*fl. 2nd half 3rd century* B.C.), which is unknown, but reported to be better. However, Archimedes himself did better than $\tfrac{22}{7}$. His lower bound for π, namely $3\tfrac{10}{71} = \tfrac{223}{71}$ will yield here

$$\frac{360 \cdot 60}{2\pi} \approx \frac{360 \cdot 60}{2} \cdot \frac{71}{223} \approx 3438.565\ldots \approx 3439,$$

and, if we go back to Chapter 1 and use, not the simple fractions, but the estimate 3.14103951 for π obtained by calculating the perimeter of the inscribed regular 96-gon, we obtain

$$\frac{360 \cdot 60}{2\pi} \approx \frac{360 \cdot 60}{2(3.14103)} \approx 3438.36\ldots \approx 3438.$$

The most famous estimate for π after $\tfrac{22}{7}$ is $\tfrac{355}{113}$, which very quickly yields

$$\frac{360 \cdot 60}{2\pi} \approx \frac{360 \cdot 60}{2} \cdot \frac{113}{355} \approx 3437.746\ldots \approx 3438.$$

More pertinent to our general discussion is perhaps Hipparchus's use of linear interpolation. I confess I do not remember whether we learned this in elementary school or high school, and I do not know if it is still taught there. It is a simple enough matter. A table for a unary function f is just a pair of

x	0	1	2	3	4	5
$f(x)$	0	1	4	9	16	25

TABLE 1. Values of the Function f

rows listing corresponding values of x and $f(x)$ in the various columns as in TABLE 1, above. The function f is quickly recognisable as $f(x) = x^2$ and if we wanted the value of f at, say, 2.3 we would just calculate $f(2.3) = 2.3^2 = 5.29$. If, however, we pretend we don't recognise the function, we might plot the points in the table and connect the dots as in FIGURE 1. This approximates

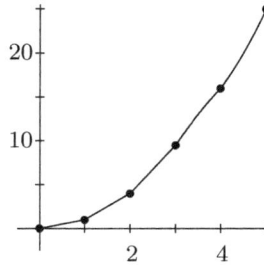

FIGURE 1. Linear Interpolation

the function on each of the segments $[n, n+1]$ for $n = 0, 1, 2, 3, 4$ by a straight line. The equation of the line on the interval $[2, 3]$ is given by

$$\frac{y - f(2)}{x - 2} = \frac{f(3) - f(2)}{3 - 2},$$

i.e.,

$$\frac{y - 4}{x - 2} = \frac{9 - 4}{1},$$

i.e.,

$$y = 4 + 5(x - 2) = 5x - 6.$$

Plugging in $x = 2.3$, we would get $y = 5(2.3) - 6 = 11.5 - 6 = 5.5$, not a particularly close approximation, but a good rough estimate. I suspect we learned this in elementary school, because we were not taught this geometrically via the equation of a line, but through solving a simple proportion:

$$\frac{2.3 - 2}{3 - 2} = \frac{y - 2^2}{3^2 - 2^2}$$

$$.3 = \frac{y - 4}{5}$$

$$y = 1.5 + 4 = 5.5.$$

The next famous trigonometric table to come after Hipparchus was Ptolemy's table of chords. Ptolemy divided the circumference of the circle into degrees, minutes, and seconds and chose a diameter of 120, the radius thus being 60. Like Hipparchus, he noted the chords of 90° and 60° are easily found, but preceded this remark with a clever determination of the sides of the inscribed regular decagon and pentagon, and thus the chords of the angles 36° and 72°. Like Hipparchus, he used formulæ for the chords of the half-angle and the supplementary angle to find more entries in his table. Unlike Hipparchus, he also had the means to calculate the chords of sums and differences of angles. Thus, he could calculate the chord of 12° = 72° − 60°, and the chord of half that or 6°, and the chord of half that or 3°, the chord of half that or $1\frac{1}{2}°$, etc.

Ptolemy's table of chords gives the chords of angles for every half degree from $\frac{1}{2}°$ to 180°. We can see how he could get the values for every $1\frac{1}{2}°$ from $1\frac{1}{2}°$ to 180°, but how do we account for the chords of $\frac{1}{2}°$ and 1°? Halving $1\frac{1}{2}°$

results in $\frac{3}{4}^\circ$, not $\frac{1}{2}^\circ$ or 1°. Ptolemy obtained a value for the chord of 1° by an unusual interpolation accompanied by a proof that the difference between the actual value and the interpolated value was negligible. It uses a lemma:

1 Lemma. *If two unequal chords are inscribed in a circle, the greater has to the less a ratio less than the arc on the greater has to the arc on the less.*

The proof of this would take us too far afield. I refer the reader to Ptolemy or Copernicus[1] for the details.

Ptolemy applied the Lemma twice, first to the pairs $\frac{3}{4}^\circ$ and 1° and then to 1° and $1\frac{1}{2}^\circ$:

$$\frac{\text{chord}\,1}{\text{chord}\,\frac{3}{4}} < \frac{1}{3/4},$$

i.e.,

$$\text{chord}\,1 < \frac{4}{3}\,\text{chord}\,\frac{3}{4},$$

and

$$\frac{\text{chord}\,1\frac{1}{2}}{\text{chord}\,1} < \frac{3/2}{1}$$

$$\frac{2}{3}\,\text{chord}\,\frac{3}{2} < \text{chord}\,1.$$

But to 3 sexagesimal places,

$$\frac{2}{3}\,\text{chord}\,\frac{3}{2} = 1^\circ 2' 50'' = \frac{4}{3}\,\text{chord}\,\frac{3}{4},$$

whence chord 1 is $1, 2, 50$ to 3 sexagesimals.

Discussing Copernicus next is placing him too early in chronological order, but his trigonometric tables are really just a mildly refined translation of Ptolemy's tables into decimal form. Ptolemy used a radius of 60, being $1,0$ in sexagesimal, and Copernicus used 100000, essentially 1 in decimal notation, but with extra 0's to avoid dealing with fractions in his calculations to 5 decimals—roughly the same order of accuracy as Ptolemy's 3 sexagesimals. The other differences are that Copernicus tabulated the sines of angles between 0° and 90°, equivalent to calculating chords, *á la* Ptolemy, between 0° and 180°. And, where Ptolemy proceeded by increments of half a degree, Copernicus went by a sixth of a degree—which would be equivalent to a third of a degree in Ptolemy's table.

The switch over from chords to sines was introduced by the Indians, who were influenced as much by Hipparchus as by Ptolemy. They did one more halving of angles than Hipparchus and constructed several tables with increments of $3\frac{3}{4}^\circ$. The *Paitāmahasiddhānta* of the early 5th century A.D. contains such a

---

[1] In the *Great Books of the Western World* edition, these are on pp. 19 - 20 and 536 - 537, respectively. The former cites the results from *The Elements* used in the proof.

sine table based on a circle of radius 3438.[2] Brahmagupta later gave a similar table for a radius 3270 in his *Brāhmasphuṭasiddhānta*[3]. Still later, Śrīpati (*fl.* 1039 - 1056) gave a similar table for a circle of radius 3415.[4]

Bhāskara's tables are particularly interesting. He used an extrapolation formula based on the second differences. To explain this, note that in most tables the differences between successive values of x are constant, in the tables mentioned thus far these differences are $\frac{1}{2}^\circ$ (Ptolemy), $\frac{1}{6}^\circ$ (Copernicus), and $3\frac{3}{4}^\circ$ (Indian). Beginning with Ptolemy and continuing through the Indians, tables would also list the differences between the values, as in TABLE 2, below, where

x	x_0	x_1	x_2	x_3	$\cdots$
f	$f(x_0)$	$f(x_1)$	$f(x_2)$	$f(x_3)$	$\cdots$
Δf	$f(x_1)-f(x_0)$	$f(x_2)-f(x_1)$	$f(x_3)-f(x_2)$	$f(x_4)-f(x_3)$	$\cdots$

TABLE 2. Values and Differences of the Function f

$\Delta f(x_i) = f(x_{i+1}) - f(x_i)$. One might also want to calculate the second differences,

$$\Delta^2 f(x_n) = \big(f(x_{n+2}) - f(x_{n+1})\big) - \big(f(x_{n+1}) - f(x_n)\big).$$

According to Bhāskara, if the differences $x_{n+1} - x_n$ all equal $3\frac{3}{4}^\circ$, then

$$\Delta^2 f(x_n) = -\frac{\sin(x_{n+1})}{225}. \tag{1}$$

Taking three successive values of x_n with a constant difference of h, say $a - h, a, a + h$, one has

$$\sin(a + h) - \sin a = \sin a \cos h + \sin h \cos a - \sin a$$
$$\sin a - \sin(a - h) = \sin a - (\sin a \cos h - \sin h \cos a).$$

Lining up terms and subtracting, this yields

---

[2] Information on Indian trigonometry is incorporated in David Pingree's essay "History of Mathematical Astronomy in India", in: *Dictionary of Scientific Biography* 15 (1978). The table in question appears on p. 559.

[3] *Ibid.*, p. 570. Pingree calls 3270 an "odd value" and, indeed, it is not clear what consideration governed the choice.

[4] *Ibid.*, p. 582. Pingree calls 3415 an "unusual value". However, it can be explained by choosing $\sqrt{10}$ as one's approximation to π:

$$\frac{360 \cdot 60}{2\sqrt{10}} \approx 3415.259873,$$

as Pingree himself had already noted on p. 579. Moreover, the use of $\sqrt{10}$ as an estimate for π was common in Indian astronomy. It was used, for example in the *Paitāmaha* and is the value used by Brahmagupta and Śrīpati, so 3415 is hardly an "unusual value", at least not in the same sense that Brahmagupta's 3270 is "odd".

$$\Delta^2 \sin(a - h) = \big(\sin(a + h) - \sin a\big) - \big(\sin a - \sin(a - h)\big)$$
$$= 2 \sin a \cos h - 2 \sin a = -2(1 - \cos h) \sin a$$
$$= -2(\text{verssin } h) \sin a,$$

using the *versed sine*, an auxiliary trigonometric function introduced by the Indians. Bhāskara now used the approximations

$$\sin\left(3\frac{3}{4}^\circ\right) = \frac{100}{1529}, \quad \cos\left(3\frac{3}{4}^\circ\right) = \frac{466}{467}, \tag{2}$$

both accurate to about one ten-millionth. The latter gives us

$$\Delta^2 \sin(a - h) = -2\left(1 - \frac{466}{467}\right) \sin a = -\frac{2}{467} \sin a = -\frac{\sin a}{233.5}. \tag{3}$$

Probably for computational simplicity, Bhāskara replaced 233.5 by 225 thus yielding a factor

$$-\frac{1}{225} = -\frac{4}{900} = -.00\overline{4},$$

agreeing to 3 decimals with

$$-\frac{1}{233.5} = -.00428\ldots\,{}^5$$

Using (2) to start with and a double-angle formula to calculate $\sin 7^\circ$ and $\cos 7^\circ$, the standard Indian table can now be generated fairly easily: One imagines the table to be constructed to be of the form in TABLE 3, below. Here

x	x_0	x_1	x_2	$\ldots$
$f(x)$	$f(x_0)$	$f(x_1)$	$f(x_2)$	$\ldots$
$\Delta f(x)$	$\Delta f(x_0)$	$\Delta f(x_1)$	$\Delta f(x_2)$	$\ldots$
$\Delta^2 f(x)$	$\Delta^2 f(x_0)$	$\Delta^2 f(x_1)$	$\Delta^2 f(x_2)$	$\ldots$

TABLE 3.

$f(x_0) = \sin 3\frac{1}{2}^\circ$ and $f(x_1) = \sin 7^\circ$ are known, whence so is

$$\Delta f(x_0) = f(x_1) - f(x_0),$$

and one takes

---

[5] In sexagesimals,
$$\frac{1}{233.5} = \frac{15}{60^2} + \frac{25}{60^2} + \ldots, \quad \frac{1}{225} = \frac{16}{60^2}.$$
Note, in this respect, that
$$\frac{1}{240} = \frac{15}{60^2}$$
would seem to be an even more convenient choice.

$$\Delta^2 f(x_0) = -\frac{f(x_1)}{225},$$

in accordance with (3), thus completing the first column. Given the values $f(x_n), \Delta f(x_n), \Delta^2 f(x_n)$, one has

$$f(x_{n+1}) = f(x_n) + \Delta f(x_n)$$
$$\Delta f(x_{n+1}) = \Delta f(x_n) + \Delta^2 f(x_n)$$
$$\Delta^2 f(x_{n+1}) = -f(x_{n+1})/225,$$

i.e., the elements of the next column. Thus one can complete the table one column at a time.

Bhāskara also gave improved estimates for $\sin 1°$ and $\cos 1°$:

$$\sin 1° = \frac{10}{573}, \quad \cos 1° = \frac{6568}{6569},$$

the former correct to 6 decimals and the latter to 7, and constructed tables with an increment of $1°$. Neither of my sources explains how he derived these values.[6]

The great age of Arab mathematics overlapped that of Indian mathematics, with the Arabs inheriting and combining both traditions and possibly influencing later Indian developments. Their tables show the influence of Ptolemy, but calculate sines instead of chords. Ḥabash al-Ḥāsib (*died* 864 - 874) computed a table of sines for every quarter degree from $0°$ to $90°$ and tangents for every half degree from the same range.

Abū al-Ḥasan ibn Yūnus (*died* 1009) is worth commenting on for his improvement on Ptolemy's estimate for the sine of $1°$:

$$\sin 1° \approx \frac{1}{3}\frac{8}{9}\sin\left(\frac{9°}{8}\right) + \frac{2}{3}\frac{16}{15}\sin\left(\frac{15°}{16}\right).$$

Note here that $\frac{9}{8}°$ and $\frac{15}{16}°$ are obtained from $18°$ and $30°$, respectively, by successive halving, and their sines can be found from those of $18°$ and $30°$ via repeated use of the half-angle formula. Hankel[7] offers the following justification: By Lemma 1

$$\frac{\sin(\alpha + \delta)}{\sin\alpha} < \frac{\alpha + \delta}{\alpha}$$

for small α, δ. In particular, for $\alpha = 1°$ and $\delta = -\frac{1}{16}°, \frac{2}{16}°$ this yields

$$\frac{16}{15}\sin\left(\frac{15°}{16}\right) > \sin 1° > \frac{16}{18}\sin\left(\frac{18°}{16}\right).$$

---

[6] These are Pingree, *op. cit.*, and Hermann Hankel, *Zur Geschichte der Mathematik in Alterthum und Mittelalter*, Verlag von B.G. Teubner, Leipzig, 1874, pp. 217 - 218. My information on Bhāskara derives from Hankel, the rest on Indian trigonometric tables from Pingree's essay. And most of the rest of my information on trigonometric tables was mined from the articles in the *Dictionary of Scientific Biography*.

[7] *Op. cit.*, p. 288, footnote *.

One would now linearly interpolate between these extremes:

$$\frac{\sin 1 - \frac{16}{15}\sin\frac{15}{16}}{\frac{16}{18}\sin\frac{18}{16} - \frac{16}{15}\sin\frac{15}{16}} \approx \frac{1 - \frac{15}{16}}{\frac{18}{16} - \frac{15}{16}} = \frac{1}{3}$$

$$\sin 1 - \frac{16}{15}\sin\frac{15}{16} \approx \frac{1}{3}\frac{16}{18}\sin\frac{18}{16} - \frac{1}{3}\frac{16}{15}\sin\frac{15}{16}$$

$$\sin 1 \approx \frac{1}{3}\frac{8}{9}\sin\frac{9}{8} + \frac{2}{3}\frac{16}{15}\sin\frac{15}{16}$$

$$\approx .017423995,$$

with an error of approximately $.0000000069 = 6.9 \times 10^{-9}$. The result of a simple linear interpolation of sines,

$$\frac{\sin 1 - \sin\frac{15}{16}}{\sin\frac{18}{16} - \sin\frac{15}{16}} = \frac{1}{3},$$

yields

$$\sin 1 \approx \frac{1}{3}\sin\frac{9}{8} + \frac{2}{3}\sin\frac{15}{16} \approx .0174523852,$$

with a slightly larger error $.0000000212 = 2.12 \times 10^{-8}$.

Ibn Yūnus gave a table of sines at increments of $1°$, but apparently intended to extend this to increments of $\frac{1}{2}°$ as he gave instructions on how to interpolate values in a table based on half degree increments[8].

About the same time Abū'l-Wafā (940 - 997 *or* 998) used a similar modification of Ptolemy's trick to obtain mildly greater accuracy, obtaining

$$\sin\frac{1}{2}^{\circ} \approx \frac{31}{60} + \frac{24}{60^2} + \frac{55}{60^3} + \frac{54}{60^4} + \frac{55}{60^5}$$

as compared to a more correct value

$$\sin\frac{1}{2}^{\circ} \approx \frac{31}{60} + \frac{24}{60^2} + \frac{55}{60^3} + \frac{54}{60^4} + \frac{0}{60^5},$$

thus an error of only slightly less than $1/60^4$.

Another great Arab astronomer, Abū Rayḥān al-Bīrūnī (973 - *after* 1050) calculated $\sin 1°$ to 5 sexagesimals and computed tables described in the *Dictionary of Scientific Biography* as "more extensive and precise than preceding or contemporary tables". But the great computational achievement of Arabic mathematics was made by Ghiyāth al-Dīn Jamshīd Mas'ūd al-Kāshī (*died* 1429), who calculated π to 9 sexagesimals and $\sin 1°$ to 10 sexagesimals:

$$1, 2, 49, 43, 11, 14, 44, 16, 26, 17.$$

The task of filling a table falls into two parts—calculating exactly, or, at least, to the desired degree of accuracy, a lot of values, and interpolating additional values based on these. The sines of the angles $1\frac{1}{2}°, 3°, 4\frac{1}{2}°, \ldots$ and

---

[8] Cf. J.L. Berggren's *Episodes in the Mathematics of Medieval Islam*, Springer-Verlag, New York, 1986, pp. 149 - 151 for details.

their halves, and their halves, ... could all be calculated to any desired degree of accuracy by means of the usual arithmetic operations and the operation of taking square roots. The sine of 1° cannot be obtained by such operations, as first demonstrated by Pierre Wantzel (1814 - 1848) in the 19th century[9]. It is, however, the solution to a cubic equation and can be approximated as closely as desired by techniques similar to those used for finding square roots. This is what al-Kāshī did[10]. That this was a breakthrough is seen by considering that Ptolemy's interpolation,

$$\sin 1° \approx 1°2'50'',$$

had to be modified when ibn Yūnus improved the estimate to obtain

$$\sin 1° \approx 1°2'49''43'''28^{(iv)}$$

and Abū'l-Wafā had to modify it again in finding $\sin \frac{1}{2}°$. Al-Kāshī's procedure merely replaces the constant term of a polynomial by a more accurate approximation to $\sin 3°$ to work with.

The cubic equation in question is easy to derive. If θ is any angle, one can use the Addition Formula for Sines a couple of times:

$$\begin{aligned}
\sin(3\theta) &= \sin(2\theta + \theta) = \sin 2\theta \cos \theta + \sin \theta \cos 2\theta \\
&= (2\sin\theta\cos\theta)\cos\theta + \sin\theta(\cos^2\theta - \sin^2\theta) \\
&= 2\sin\theta(1 - \sin^2\theta) + \sin\theta(1 - \sin^2\theta - \sin^2\theta) \\
&= 2\sin\theta - 2\sin^3\theta + \sin\theta - 2\sin^3\theta \\
&= 3\sin\theta - 4\sin^3\theta,
\end{aligned}$$

and, if $\theta = 1°$, $\sin\theta$ is a solution to the equation

$$3'8''24'''\ldots = 3X - 4X^3,$$

or, in modern decimal notation,

$$.05233\ldots = 3X - 4X^3,$$

From this latter equation, we obtain

$$3x = 4x^3 + .05233\ldots$$

$$x = \frac{4x^3 + .05233\ldots}{3} = \frac{x^3 + .0130839\ldots}{.75}.$$

From earlier work, he knew the first few sexagesimals of $\sin 1°$. We, of course would know the first few decimals, say .017. If we suppose $x = .017abc\ldots$, we could plug this into the last equation:

---

[9] A proof of this impossibility, accompanied by a bit of its history, can be found in my *History of Mathematics; A Supplement*, Springer-Verlag, New York, 2008.

[10] I adapt the presentation from Berggren, *op.cit.*, pp. 151 - 153.

$$.017abc\ldots = \frac{(.017abc\ldots)^3 + .0130839\ldots}{.75}$$

and evaluate the right-hand side:

$$.017abc\ldots \approx \frac{.000004913\ldots + .0130839\ldots}{.75}$$

$$\approx .0174517507,$$

suggesting 4 as the next decimal a.[11] We could now accept the 4 and try again:

$$.0174bc\ldots = \frac{(.0174bc\ldots)^3 + .0130839\ldots}{.75}.$$

In effect, al-Kāshī's solution is an iterative one: Define

$$f(x) = \frac{x^3 + .0130839\ldots}{.75}, \quad x_0 = .017, \quad x_1 = f(x_0), \quad x_2 = f(x_1), \quad \ldots$$

Using methods of the Calculus, one easily shows the sequence $x_0, x_1, x_2, \ldots$ to converge to a limiting value.

Like al-Kāshī, Ulugh Beg (1394 - 1449) used a cubic equation to determine $\sin 1°$, and then went on to construct a very accurate table of sines for every minute from $0°$ to $45°$ and for every 5 minutes from $45°$ to $90°$.

European trigonometic tables made their first appearance in mediæval times before the work of al-Kāshī or Ulugh Beg. It followed the Arabic tradition. One of the earliest works was the Hebrew text *Ḥibbūr ha-meshīḥah we-ha-tisboret* (1116) of Abraham bar Ḥiyya (*fl. before* 1136), translated into Latin as *Liber embadorum* by Plato of Tivoli in 1145, the same year as Robert of Chester's translation of al-Khwārizmī's (*before* 800 - *after* 847) algebra book. Abraham bar Ḥiyya was a Spanish Jew living in Spain, where, in addition to writing his own works in Hebrew, he translated Arabic works into Spanish. Arabic scholars called him Ṣāhib al-Shurta, which was later Latinised to Savasorda and means "Elder of the Royal Court". The *Ḥibbūr...*, or *Treatise on Mensuration and Calculation*, was a practical geometry including a table of *arcchords*, i.e., a table

---

[11] We can analyse the error quite simply: Write $x = 17 \cdot 10^{-3} + \varepsilon$, where $\varepsilon < 10^{-3}$. Then

$$x^3 = 17^3 \cdot 10^{-9} + 3 \cdot 17^2 \cdot 10^{-6} \cdot \varepsilon + 3 \cdot 17 \cdot 10^{-3} \cdot \varepsilon^2 + \varepsilon^3$$
$$= 4913 \cdot 10^{-9} + \delta,$$

where

$$\delta < 867 \cdot 10^{-9} + 51 \cdot 10^{-9} + 10^{-9}$$
$$< 919 \cdot 10^{-9} < 10^3 \cdot 10^{-9} = 10^{-6}.$$

So the error in the numerator of the fraction is less than 10^{-6} and $10^{-6}/.75 \approx .0000013$, meaning, in fact, that we can take both 4 and 5 as the next two digits a and b.

giving the angle in terms of the chord. He assumed $\frac{22}{7}$ as a value for π and a circle of radius 14. He used the table to find lengths of arcs on his way toward finding the areas of segments of circles. As tables go, it is not particularly impressive, but the work does offer "one of the earliest uses of a trigonometric table to measure quantities on earth, rather than in the heavens. But note that trigonometry is used to measure pieces of circles rather than triangles"[12].

The next century or so saw a number of practical geometries, perhaps most importantly Leonardo of Pisa's *Practica geometriæ* (1220), a work strongly influenced by that of Abraham bar Ḥiyya. Leonardo, however, gave a direct table of chords based on $\pi = \frac{22}{7}$, but now with a diameter of 21, yielding a semi-circumference of 66, and featuring chords for angles from 1 to 66 parts of the circle. He did not give an arcchord table for the crucial geometric applications, but did give detailed instructions on how to find the arcchords using the chord table by interpolation.

Mediæval European trigonometric tables are not particularly innovative or noteworthy, though one should probably mention Rabbi Levi Ben Gerson (1288 - 1344) whose *De sinibus et arcubus* (1343) calculated chords, sines, versed sines, and cosines "with great precision" according to the *Dictionary of Scientific Biography*. The Renaissance and Enlightenment would change the situation as Europe once again took centre stage in mathematics.

One of the two most important figures in trigonometry in the 15th century was Johannes Müller (1436 - *c.* 1476), who was born near Königsberg, a Prussian city that pops up every so often in the history of mathematics[13]. When he enrolled at the university in Vienna in 1450, he signed himself Johannes Molitoris de Künigsperg, but he is best known under a Latin form of his name, Johannes Regiomontanus. It was his work *De Triangulis Omnimodus [On triangles of every kind]*, written in 1463, but first published in 1533, that offered the "first printed systematization of [trigonometry]... as a branch of mathematics independent of astronomy"[14]. The tables in this book were sexagesimal and based on a circle of radius 60000. His next important work was his TABLES OF DIRECTIONS, written in 1467 and first published in 1490. At one point he declared that the use of decimals and a radius of 100000 were a better choice than his earlier 60000, and the tangent table he included was based on such a

---

[12] Victor J. Katz, *A History of Mathematics; An Introduction*, HarperCollins College Publishers, New York, 1993, p. 271. Katz's textbook offers nice discussions of the history of trigonometry, which is not really the topic of the present chapter, in separate sections in chapters 4 (Mathematical Methods in Hellenistic Times), 6 (Medieval China and India), 7 (The Mathematics of Islam), 8 (Mathematics in Medieval Europe), and 10 (Mathematical Methods in the Renaissance).

[13] There is most famously the problem of the seven bridges of Königsberg discussed by Euler. Immanuel Kant (1724 - 1804), not a mathematician, but a philosopher whose ideas about mathematics were often taken seriously, lived in Königsberg. David Hilbert (1862 - 1943), the great mathematician of the early 20th century, also came from there. And the first public announcement by Kurt Gödel (1906 - 1978) of his famous First Incompleteness Theorem was made at a conference in Königsberg.

[14] *Dictionary of Scientific Biography*, vol. 11, p. 350.

radius. Before completing this work he had calculated a sexagesimal table of sines for every minute of arc for a circle of radius 6000000. In 1468, he computed a decimal sine table using 10000000 as radius. Both tables, together with an essay *Construction of Sine Tables*, were first published in 1541.

In my trigonometrocentric discussion, I should not overlook another set of tables by Regiomontanus. His *Ephemerides*, published in 1474, was the first printed ephemerides and a copy accompanied Columbus (1451 - 1506) on his fourth voyage. Indeed, it was its prediction of a lunar eclipse on 29 February 1504 that Columbus used "to frighten the hostile Indians in Jamaica into submission"[15].

This last anecdote brings us to an important point. One of the stimuli at this time for more—and more accurate—tables in Europe was the growing age of exploration. Sailing within a small, enclosed body of water, like the Mediterranean Sea, gave way to voyages across the open ocean and a greater need for tables to aid in navigation arose. The table-makers responded.

Abraham bar Samuel bar Abraham Zacuto (*c.* 1450 - *c.* 1522) was a Spanish Jew, relocated to Portugal following the expulsion of Jews from Spain. The Portuguese king quickly realised the value of this new acquisition and it was Zacuto who prepared the solar tables for the voyage (1497 - 1500) of Vasco da Gama (1449 - 1526) around Africa. This was done on the basis of Zacuto's *Almanach perpetuum*, first published in 1496. This work served as the basis of all solar tables for the next 40 years at which time new tables by Pedro Nuñez Salaciense (1502 - 1578) appeared.

Returning to trigonometric tables, Regiomontanus was followed by Georg Joachim Rheticus (1514 - 1574), famous for having published the first authorised announcement of the Copernican system in 1540 and having arranged its publication in Nürnberg in 1543. A year earlier, in 1542, he published *De lateribus et angulis triangulorum*, the trigonometric section of Copernicus's *De revolutionibus*. The included sine tables, based on a radius of 10000000 as opposed to the Copernican 100000, were more extensive, sines being calculated at every minute of arc as opposed to the ten minute intervals of Copernicus.

In 1551 Rheticus published the *Canon of the Doctrine of Triangles*, the first tables to include "all 6 trigonometric functions, including the first extensive table of tangents and the first printed table of secants"[16]. He was at work calculating an even more extensive table of tangents when he died, and the work was passed on to Lucius Valentinus Otho, who, with much difficulty, some royal patronage, and computing assistants, finished the task of calculating a table of sines from 0° to 90° using an increment of 45 seconds and a circle of radius 10^{15}. It was published in 1596 as *Opus Palatinum de triangulus*.

The values from 83° to 90° in Otho's *Opus* contained serious errors in the tangent and secant tables. These were corrected by Bartholomeo Pitiscus (1561 - 1663), whose *Thesaurus Mathematicis* (1613) included two tables by Rheticus,

1. sines for every 10 seconds to 15 decimals,

---

[15] *Ibid.*, p. 351.
[16] *Ibid.*, p. 397.

2. sines for every second from $0°$ to $1°$ and from $89°$ to $90°$, also to 15 decimals,

as well as two sections of his own work including sines to 22 decimals for every 10th, 30th, and 50th second in the first 30 minutes.

It boggles the mind to think of the immense effort involved in all that calculation done by hand. There always remained the work of completing the table by filling in more values, extending the precision by adding more decimals—which would only be of practical use if the the increments of the angles at which they were calculated were sufficiently small.[17] Even with the tables frozen there was more work to do as one could add new kinds of tables. For example, in his *De motu gravium* of 1641, Evangelista Torricelli (1608 - 1642) accompanied his basic sine table (for every degree from $0°$ to $90°$) by three more trigonometric tables—$\sin 2\alpha, \sin^2 \alpha, \frac{1}{2}\tan\alpha$—and by a table giving the angle of inclination needed to achieve a given distance a projectile is to go when the maximum range is known. Table-makers always had more accurate, more extensive, and newer kinds of tables to make; and consumers of tables always had ancillary calculations to perform. Even something as straightforward as linear interpolation involved addition, subtraction, multiplication, and division, and, whereas addition and subtraction are simple enough, multiplication and division are laborious, time-consuming tasks. But there were a lot of trigonometric tables around and soon there was *prosthaphairesis*.

Prosthaphairesis derives its name from two Greek words πρόσθεσις [addition] and ἀφαίρεσις [subtraction] and is a manifestation of the Addition and Subtraction Formulæ for Sines and Cosines. Suppose one wants to multiply two multidigit numbers, say 61620 and 45318. First one might write them as

$$.61620 \times 10^5 \quad\text{and}\quad .45318 \times 10^5,$$

respectively, and note that their product will be

$$.61620 \times .45318 \times 10^{10},$$

the difficult part being the first product. Now, looking the numbers up in a table of sines, one finds

$$.61620 = \sin 38.039°$$
$$.45318 = \sin 26.948°.$$

From

$$\cos(\alpha + \beta) = \cos\alpha\cos\beta - \sin\alpha\sin\beta$$
$$\cos(\alpha - \beta) = \cos\alpha\cos\beta + \sin\alpha\sin\beta$$

---

[17] Consider the tables of square roots given to 50 decimals in the last chapter. Suppose you wanted to know $\sqrt{2.5}$ and interpolated a value in the table. How many decimals of your estimate are you confident in? 50? 25? 2? 1? For ordinary linear interpolation, using all 50 decimals in $\sqrt{2}$ and $\sqrt{3}$, the result is only correct to 1 decimal place.

one derives

$$\cos(\alpha + \beta) - \cos(\alpha - \beta) = -2\sin\alpha\sin\beta,$$

i.e.,

$$\sin\alpha\sin\beta = \frac{\cos(\alpha - \beta) - \cos(\alpha + \beta)}{2}.$$

Thus

$$.61620 \times .45318 = (\sin 38.039°)(\sin 26.948°)$$
$$= \frac{1}{2}(\cos 11.091° - \cos 64.987°)$$
$$= \frac{1}{2}(.9813219765 - .4228229495)$$
$$= .279249516,$$

and multiplying by 10^{10} yields

$$61620 \cdot 45318 = 2792495160.$$

This is indeed correct and we obtained it using only addition and trigonometric functions for which plenty of tables existed. (Needless to say, I cheated and used the calculator for the trigonometric and inverse trigonometric functions.) The accuracy of the final result depends, of course, on the extent and accuracy of one's trigonometric tables.

Prosthaphairesis was largely the work of Paul Wittich (1555(?) - 1587), an assistant to the famed astronomer Tycho Brahe (1546 - 1601), who was not overly pleased that information about the method leaked to the rest of Europe from his observatory. It didn't matter, for in very short order John Napier, the Laird of Merchiston, (1550 - 1617) and Joost Bürgi (1552 - 1632) had independently invented logarithms. While Bürgi's invention was an outgrowth of his work on improving prosthaphairesis, Napier was well on his way to inventing logarithms when he first heard of the trigonometric practice. Bürgi seems to have discovered the logarithms first, but didn't publish until 1620, half a decade after Napier, and he had little impact on subsequent developments. Napier wrote two works on his logarithms: *Mirifici Logarithmorum Canonis Descriptio* [*Description of the admirable table of logarithms*[18]] (1614) and *Mirifici Logarithmorum Canonis Constructio* [*Construction of the admirable table of logarithms*] (1619), the latter published posthumously. The first of these books was translated into English in 1616, and its title page is almost a monograph itself:

A

DESCRIPTION

OF THE ADMIRABLE

TABLE OF LOGA-

RITHMES:

---

[18] The word "canon" by this time had come to mean "table".

WITH
A DECLARATION OF
THE MOST PLENTIFVL, EASY,
and speedy vse therof in both kindes
of Trigonometrie, as also in all
Mathematicall calculations,
INVENTED AND PVBLI-
SHED IN LATIN BY THAT
Honorable L IOHN NEPAIR, Ba-
ron of *Marchiston,* and translated into
English by the late learned and
famous Mathematician
Edward Wright.
With an Addition of an Instrumentall Table
to finde the part proportionall, inuented by
the Translator, and described in the end
of the Booke by HENRY BRIGS
Geometry-reader at Gresham-
house in London.
All perused and approued by the Author,& pub-
lished since the death of the Translator.[19]

A perusal of the English edition[20] reveals almost a different universe. It starts with a dedication to the London East India Company (which had commissioned Wright to translate the text) written by Samuel Wright, son of the translator Edward Wright (1561 - 1615), expressing his belief that the work will be of use to mariners:

> I am the bolder thus to do, in regard it is not vnknowne to many men, that my said father spent a great part of his time in study of the Art of Nauigation, and had gathered much vnderstanding by his owne practise in some voyages to sea with the right honourable the Earle of Cumberland deceased: whereupon he published a painful worke discouering errours committed by Mariners in that Art, with corrections and ready wayes for reformation therof. So that I thinke it is out of doubt, that his iudgement therein was great. And seeing hee not only gaue much commendation of this worke (and often in my hearing) as of very great vse for Mariners: but also to help the want of those that could not vnderstand it in Latine, translated the same into English, and added thereto an instrumentall Table to finde the part proportional, wherof also the noble Author approued well. I doubt not but it is apparent

---

[19] I have been fairly faithful to the original orthography, ignoring only the differential sizes of the type and the archaic "s" that looks almost like an "f" in upright text and like an integral sign in italic.

[20] I have a facsimile edition published jointly in 1969 by Da Capo Press, New York, and Theatrum Orbis Terrarum Ltd., Amsterdam. The book is also available online.

enough that he esteemed of it, and intended to haue recommended it as a booke of more than ordinary worth, especially to Sea-men.[21]

There follows Napier's own dedication to Prince Charles, "onely sonne of the high and mightie James", then two prefaces, one by Napier and one by Henry Briggs (1561 - 1630) whose name is there rendered "Brigges", and then two poems praising the work by John Davies of Hereford and Ri. Lever, respectively. After this seeming excess, the only things standing between the reader and the main body of the book are now two pages, one giving a graphical tree-like outline of the coming contents, and one a short list of errata to the sine table.

While under the inspiration of my own perusal of the description of the admirable table of logarithms, I shall temporarily take leave to write my own dedicatory poem in praise of the current work. While I'm away, the reader might like to ponder Napier's preface:

> Seeing there is nothing (right well beloued Students in the Mathematickes) that is so troublesome to Mathematicall practise, nor that doth more molest and hinder Calculators, then the Multiplications, Diuisions, square and cubical Extractions of great numbers, which besides the tedious expence of time, are for the most part subiect to many slippery errors. I began therefore to consider in my minde, by what certaine and ready Art I might remoue those hindrances. And hauing thought vpon many things to this purpose, I found at length some excellent briefe rules to be treated of (perhaps) hereafter. But amongst all, none more profitable then this, which together with the hard and tedious Multiplications, Diuisions, and Extraction of rootes, doth also cast away from the worke it selfe, euen the very numbers themselues that are to be multiplied, diuided and resolued into rootes, and putteth other numbers in their place, which performe as much as they can do, onely by Addition and Subtraction, Diuision by two, or Diuision by three: which secret inuention, being (as all other good things are) so much the better as it shall be more common, I thought good heretofore to set forth in Latine for the publique vse of Mathematicians... [22]

What? Finished already. All I have got so far of my self-praising, dedicatory poem is

> Just as roses are red and violets are blue,
> Every word in this book is sure to be true...

I guess I will have to finish it later.

If one continues past the preface by Briggs and the poems, one will finally come to Napier's description of his logarithms. This description, other than its examples and explanation of the use of the tables, is not as clear as that of

---

[21] *Ibid.*, pp. A3. (*N.B.* the two facing pages have but one page number.)

[22] *Ibid.*, pp. A5 - A6.

the *Constructio*[23], or works derived therefrom[24]. Napier's construction is made more complex by the need to reduce the amount of labour required to construct the tables[25], but the basic idea is simple. One starts with a large number n, and calculates successively

$$n, \quad n\left(1 - \frac{1}{n}\right), \quad n\left(1 - \frac{1}{n}\right)^2, \quad \ldots \tag{4}$$

Choosing n to be a power of 10—Napier chooses $n = 10^7$— this is actually fairly easy: multiplication and division by n consist merely of shifting decimal points and the only real computational step is subtraction. For,

$$n\left(1 - \frac{1}{n}\right)^{k+1} = n\left(1 - \frac{1}{n}\right)^k - \frac{1}{n}\left(n\left(1 - \frac{1}{n}\right)^k\right),$$

i.e., the sequence (4) for $n = 10^7$ satisfies the recurrence

$$x_0 = 10^7, \quad x_{k+1} = x_k - \frac{1}{10^7}x_k.$$

Dragging out the calculator, putting it into Sequential mode, then opening the function editor and entering

nMin $= 0$
u$(n) = u(n-1)-1/(10\wedge 7)u(n-1)$
u$(n$Min$) = \{10000000\}$,

and finally exiting the function editor and entering

u(seq$(n, n, 0, 15))\rightarrow$L$_1$ or u$(0, 15)\rightarrow$L$_1$

in the main window, will produce a list, which I have collated into TABLE 4, below. For quite a while the list agrees with v$(n) = 10^7 - n$. In fact, using greater precision, Napier calculated[26]

$$u(100) = 9999900.000490,$$

which differs from v$(100) = 10000000 - 100$ by less than 5 ten-thousandths. Napier's logarithm of a positive number x less than 10000000, which we may write NLog x, is then the value in the left column corresponding to the number's appearance on the right. Thus, for example,

---

[23] This has also been translated into English, by William Rae MacDonald in 1888 as *The Construction of the wonderful Canon of Logarithms*. The translation has been digitised and can be downloaded free online.

[24] A good example is Herman H. Goldstine, *A History of Numerical Analysis from the 16th through the 19th Century*, Springer-Verlag, New York, 1977, pp. 2 - 13. Goldstine not only describes Napier's procedure, but explains it in modern terms as well.

[25] Cf. *Exercise* 2, below.

[26] *Ibid.*, p. 6.

k	$u(k)$	k	$u(k)$
0	10000000	8	9999992
1	9999999	9	9999991
2	9999998	10	9999990
3	9999997	11	9999989
4	9999996	12	9999988
5	9999995	13	9999987
6	9999994	14	9999986
7	9999993	15	9999985

TABLE 4.

$$\text{NLog } 10^7 = 0.$$

If a number does not appear on the right, one obtains Napier's logarithm for it by interpolating between the logarithms of the nearest surrounding entries.

If we now let $a = \text{NLog } x$ and $b = \text{NLog } y$, then

$$xy = 10^7 \left(1 - \frac{1}{10^7}\right)^a \cdot 10^7 \left(1 - \frac{1}{10^7}\right)^b$$

$$= 10^7 \cdot 10^7 \left(1 - \frac{1}{10^7}\right)^{a+b}.$$

Writing

$$10^7 = \left(1 - \frac{1}{10^7}\right)^c, \tag{5}$$

we have

$$xy = 10^7 \left(1 - \frac{1}{10^7}\right)^{a+b+c},$$

and

$$\text{NLog}(xy) = \text{NLog } x + \text{NLog } y + c,$$

not quite the familiar

$$\log(xy) = \log x + \log y.$$

Now (5) implies

$$1 = 10^7 \left(1 - \frac{1}{10^7}\right)^{-c}, \tag{6}$$

whence $-c = \text{NLog } 1$, i.e., $c = -\text{NLog } 1$ and

$$\text{NLog}(xy) = \text{NLog } x + \text{NLog } y - \text{NLog } 1. \tag{7}$$

In like manner, or by algebraic manipulation of (7), one obtains

$$\text{NLog}(x/y) = \text{NLog } x - \text{NLog } y + \text{NLog } 1. \tag{8}$$

Thus, the crucial, almost defining properties of logarithms as we know them today, namely

$$\log(xy) = \log x + \log y, \quad \log(x/y) = \log x - \log y,$$

do not quite hold for Napier's logarithms, but require a correction term, namely NLog 1. For Napier's purpose, this was not a serious problem as NLog 1 did not appear in his initial applications. Nonetheless, let me calculate NLog 1 anyway. Taking modern logarithms of the two sides of (6), we have

$$0 = \log 1 = \log 10^7 - c \log \left(1 - \frac{1}{10^7}\right),$$

from which we conclude

$$c = \frac{\log 10^7}{\log \left(1 - \frac{1}{10^7}\right)} = -161180948.5,$$

whence

$$\text{NLog } 1 = 161180948.5. \tag{9}$$

As I said, NLog 1 does not occur in Napier's initial applications or in the *Descriptio* at all. The reason for this is that in that work he is not interested in using logarithms directly to find products, but rather for solving proportions where NLog 1 does not pop up. The first proposition of Chapter 2 of the first book of the *Descriptio* reads

> The Logarithmes of Proportionall numbers and quantities are equally differing.[27]

This means, if $x/y = z/w$, then

$$\text{NLog } x - \text{NLog } y = \text{NLog } z - \text{NLog } w,$$

which follows[28] from (8):

$$\text{NLog } x - \text{NLog } y = \text{NLog}(x/y) - \text{NLog } 1$$
$$= \text{NLog}(z/w) - \text{NLog } 1 = \text{NLog } z - \text{NLog } w.$$

His second proposition reads

> Of the Logarithmes of three proportionals, the double of the second or meane, made lesse by the first, is equall to the third.[29]

---

[27] Napier, *Description*, p. 7.
[28] In the *Descriptio*, Napier's description of logarithms and proof of this are geometric, rather than algebraic.
[29] *Ibid.*

That is, if $x/y = y/z$, we have

$$\mathrm{NLog}\,x - \mathrm{NLog}\,y = \mathrm{NLog}\,y - \mathrm{NLog}\,z,$$

whence

$$2\,\mathrm{NLog}\,y - \mathrm{NLog}\,x = \mathrm{NLog}\,z. \qquad (10)$$

The use of this proposition is illustrated in Chapter 5, where he sets out to solve a proportion. Before presenting his example, however, we need to explain a few things.

The *Descriptio* does not contain logarithmic tables *per se*, but contains instead a logarithmic sine table, or collection of such tables, one for each degree from $0°$ to $44°$. Each such table covers two pages and has 7 columns. The first column consists of minutes to be added to the number n of degrees of the individual table and these number from 0 to 60. The seventh column contains the corresponding minutes to add to $90 - n - 1$ to obtain the complementary angle. Thus, for example, one has $24°12'$ paired with $65°48'$. The second and sixth columns contain the sines of the two complementary angles, or, if one prefers, the sine and cosine of the angle in the first column. The logarithms of these sines occupy the third and fifth columns, and their difference the central, fourth column. Napier refers to the entry in this column as the logarithm of the tangent, although, by (8) one must add NLog 1 to obtain the actual logarithm of the tangent. Likewise the negated values of the third and fifth columns are, up to NLog 1, the logarithms of the cosecant and secant, respectively.

Now all of what I've just said is true provided one realise that the sine and cosine at that time were not fixed, but were still the lengths of various line segments in a circle of some given radius. In constructing his tables, Napier took the radius to be 10^7 and the tables in the original Latin edition of the *Descriptio* do the same. Thus, for example, we read in Victor Katz's textbook[30] that he "present[s] one line of Napier's actual tables" and we see the line

---

[30] Katz, *op. cit.*, p. 382. One can also see a full page of a table from the Latin text reproduced on p. 138 of MacDonald's translation of the *Constructio*. A photographic copy of another page illustrates the article, "The making of logarithms" by Graham Jagger in: Martin Campbell Kelly, Mary Croarken, Raymond Flood, and Eleanor Robson, eds., *The History of Mathematical Tables: From Sumer to Spreadsheets*, Oxford University Press, Oxford, 2003. *Cf.* p. 48. This book is a collection of well-written historical articles that hang together a bit better than one would expect from the proceedings of a conference. The articles, mostly by professional historians, are strong on historical detail but don't offer much technical detail. For this latter, I have found Goldstine (*op. cit.*), Katz (*op. cit.*), the *Dictionary of Scientific Biography*, and even Jean Luc Chabert, *A History of Algorithms: From the Pebble to the Microchip*, Springer-Verlag, Berlin, 1999, more useful for my purposes. Another apparently useful reference, had I learned of it before writing the present chapter, would have been Charles Hutton, *Mathematical Tables; Containing the Common, Hyperbolic, and Logistic Logarithms; Also Sines, Tangents, Secants, and Versed Sines* (the title goes on a bit, eventually coming to) *To which is prefixed, A Large and Original History of the Discoveries and Writings Relating to These Subjects*, 5th ed., London, 1811. Google Books has several versions available for download, not all of which (e.g., the twelfth) have

$$34°40' \quad 5688011 \quad 5642242 \quad 3687872 \quad \ldots,$$

but if we consult the English translation, we read

$$34°40' \quad 568801 \quad 564224 \quad 368787 \quad \ldots$$

MacDonald verifies the obvious conjecture:

> The Table is to one place less than the Canon of 1614, but the loga-
> rithms of the sines for each minute for 89° - 90° are given in full, the
> last figure being marked off by a point. This is, I believe the earliest
> instance of the decimal point being used in a printed book.[31]

The reason for the lower precision might be a lesser need in navigation than in
astronomy for that extra digit, except for the logarithms in the region between
89° and 90° where the differences in the logarithms of the sines are particularly
small. Another reason might be the comparitive sizes of the books: whereas
the Latin edition was $7\frac{1}{2} \times 5\frac{3}{4}$ inches, the English version for mariners, who
would presumably have less shelf space and smaller working surfaces, was only
$5\frac{7}{8} \times 3\frac{3}{8}$ inches. The additional 5 digits would almost add another column for
which there was scarcely enough room.

So far as I can see, the tables of the English version, though now apparently
based on a radius of 10^6, were obtained from the Latin text by truncation or
rounding, not by recalculation.

With these pieces of information, we can present one of Napier's applications. In Chapter 5 of the first book, he sets out to solve a proportion:

> LET the first proportionall giuen, bee 1000000, and the second 707107:
> let the third be sought for, which commonly is found by multiplying
> the middle number by it selfe, & diuiding this square by the first[.] But
> we find it easilier by doubling the Log: of the middle number 346573,
> and by subtracting from this double (wch is 693147) the Logarithme of
> the first, which is 0, & there remaineth 693147, the Logarithme sought
> for, whose arch you shall finde to be 30 degrees, and the Sine adioyning
> thereto 500000, which is the proportionall number sought for.[32]

In modern terminology, Napier says that he wishes to solve the proportion

$$\frac{1000000}{707107} = \frac{707107}{x}.$$

Instead of performing a multiplication and division,

the history, which offers a wealth of detail on European developments, particularly
regarding logarithms and the methods of construction of logarithmic tables. Also
informative is H. Andoyer's survey, "Fundamental trigonometrical and logarithmic tables", which appeared in: Cargill Gilson Knott, ed., *Napier Tercentenary
Memorial Volume*, Longmans, Green and Company, London, 1915.

[31] MacDonald, *op. cit.*, pp. 145 - 146.
[32] *Description*, p. 24.

$$x = \frac{707107^2}{1000000},$$

as one usually does, he applies (10):

$$\begin{aligned} \text{NLog}\,x &= 2\,\text{NLog}\,707107 - \text{NLog}\,1000000 \\ &= 693147 - 0 \\ &= \text{NLog}(\sin 30°) \\ &= \text{NLog}\,500000, \end{aligned}$$

whence $x = 500000$.

So long as he uses his logarithms to solve proportions, the ugly correction term NLog 1 will drop out, just as (for those who know Calculus) the constant of integration disappears when one's integrals are all definite. However, for performing multiplications by translating them into additions via logarithms, Napier's logarithms had two disadvantages, one being that the term NLog 1 was nonzero, and the other being that NLog 10 was not 1. Henry Briggs suggested constructing a table in which the logarithm of the whole sine should be 0 and that of $1/10$ of the whole sine should equal the whole sine 10^{10}, that is, if we divide by the radius to normalise, he suggested

$$\log 1 = 0 \quad \log .1 = 1.$$

Napier had also been thinking along these lines and suggested one reverse directions, making the logarithm of x grow as x grows instead of diminishing. In modern terms:

$$\log 1 = 0, \quad \log 10 = 1.$$

Briggs agreed and subsequently calculated the first table of logarithms to base 10.

Before discussing future logarithmic tables, I should clarify the relation between Napier's logarithm, NLog x, and the so-called *Napierian* or *natural logarithm*, ln x, which is the logarithm to the base $e = 2.71828\ldots$ In the Calculus, one proves

$$\lim_{n\to\infty}\left(1 + \frac{y}{n}\right)^n = e^y,$$

i.e., for large values of n, the number

$$\left(1 + \frac{y}{n}\right)^n$$

is very close to e^y. Let some number x occur in Napier's geometric progression (4),

$$x = n\left(1 - \frac{1}{n}\right)^k.$$

For such x,

$$x = n\left(1 - \frac{1}{n}\right)^k = n\left(1 - \frac{1}{n}\right)^{n\cdot(k/n)}$$

$$\approx n \cdot e^{-k/n}$$

if n is sufficiently large. Then

$$\ln x \approx \ln\left(n \cdot e^{-k/n}\right) = \ln n - \frac{k}{n}\ln e \approx \ln n - \frac{k}{n},$$

whence

$$k \approx n \ln n - n \ln x.$$

In particular,

$$\text{NLog}\, x = k \approx 10^7\left(\ln 10^7 - \ln x\right) \tag{11}$$
$$\approx 10^7(16.1180956 - \ln x)$$
$$\approx 161180956 - 10^7 \ln x. \tag{12}$$

Plugging $x = 10^7$ into (11) gives 0 on the right, as was to be expected, and plugging $x = 1$ into (12) yields

$$\text{NLog}\, 1 \approx 161180956, \tag{13}$$

which is relatively close to the correct value given by (9). I should emphasise that the equalities (11) and (13) are only approximate, a *caveat* not always emphasised in histories of the subject.

2 Exercise. Assume Napier had stuck with the plan and continued calculating

$$10^7\left(1 - \frac{1}{10^7}\right)^k$$

until he got as far as 1. The number $\text{NLog}\, 1 = 161180948.5$ approximates the number of steps it would have taken him, i.e., the number of logarithms he would have had to have calculated, before finishing the table at 1.
i. Assume he could fit 8 columns, alternately labelled $x, \text{NLog}\, x$ on a page, and each column could fit 100 numbers. How many pages would his table of logarithms have run to? Assume he bound them in volumes running 500 pages each. How many volumes would that amount to?
ii. It is probably unrealistic to assume he could complete a page in 15 minutes, but assume he could do so. A 40 hour work week wouldn't indicate any passion for the work, so assume he worked 60 hours a week at calculating logarithms. To the nearest year, how long would it have taken him to complete the table? It is reported he spent about 20 years labouring over the concept and tabulations[33]. What does that tell you about his method?

To construct the table as he did, Napier had to replace the straightforward, naïve approach by a more workable one, as the above exercise makes clear. But his methods were still quite cumbersome. Briggs, in calculating his logarithms

---

[33] Goldstine, *op. cit.*, p. 3.

to base 10 anew, was much cleverer. It would require too much space to discuss here, so I refer the reader to Goldstine's exposition[34] for the details.

In 1617 Briggs published the first printed table of logarithms to base 10. Called *Logarithmorum chilias prima* after the first words of its preface, it is only a short 16 page pamphlet giving a table of logarithms of the first 1000 integers, each to 14 decimal places. He extended this in 1624 with with his publication of *Arithmetica Logarithmorum* in which he tabulated logarithms for integers from 1 to 20000 and from 90000 to 100000. In 1620 Edmund Gunter (1581 - 1626), a close friend of Briggs, published another table of decimal logarithms under the title *Canon triangulorum sive tabulæ sinuum et tangentium artificialium ad radium 10000.0000 & ad scupulum prima quadrantis* [*Trigonometrical canon, that is tables of artificial[35] sines and tangents to a radius of 100000000 for each minute of the first quadrant*]. Gunter's layout is similar to Napier's, but with some columns removed and an extra column added for the logarithm of the cotangent, i.e., the tangent of the complementary angle. The word "logarithm" does not appear in the table headings, just "Sin." and "Tan.", meaning the "artificial sine" and "artificial tangent".[36] The two works by Briggs gave tables of logarithms and Gunter's a table of logarithms of trigonometric functions. A third work by Briggs, the posthumously published *Trigonometria Britannica*, gave trigonometric functions and their logarithms. And so it continued down to the twentieth century: volume after volume of tables of logarithms and/or logarithms of trigonometric functions, to varying levels of precision and with angles given in degrees, minutes, seconds or in degrees and decimal fractions of degrees.

As I said, I do not wish to go into detail with Briggs's calculations; but I wish to cite two important innovations of his that have traditionally gone unnoticed, so much so that they had to be rediscovered and are credited to later individuals. These are the use of an infinite *power series*[37] and the Calculus of Finite Differences.

Infinite series had been around for some time, most familiarly the geometric series summing an infinite geometric progression. And power series expansions of various trigonometric functions were known to later Indian mathematicians like Madhava of Sangamagramma (*c.* 1340 -1425), but were unknown in Europe in Briggs's day and only became popular a few decades later. Briggs's

---

[34] Goldstine, *op. cit.*, pp. 13 - 20.

[35] Napier had originally called his logarithms "artificial numbers". According to Goldstine (*op. cit.*, p. 3), he compounded the word "logarithm" out of two Greek words, λόγων [ratio] and αριθμός [number].

[36] A page is reproduced on p. 64 of Jagger's article cited in footnote 30. Pages from Briggs's works can also be found therein, on pages 59 and 67, respectively. Page 68 depicts a page from *Trigonometria Britannica*, a later work of Briggs.

[37] That is, an infnite series of the form,

$$a_0 + a_1 x + a_2 x^2 + \ldots$$

If a function $f(x)$ can be represented in this form, the series is called the *power series expansion* of f.

calculation of some logarithms involved the calculation of lots of square roots and, using difference methods, he derived the series

$$(1+x)^{1/2} = 1 + \frac{1}{2}\,x - \frac{1 \cdot 1}{2 \cdot 4}\,x^2 + \frac{1 \cdot 1 \cdot 3}{2 \cdot 4 \cdot 6}\,x^3 - \frac{1 \cdot 1 \cdot 3 \cdot 5}{2 \cdot 4 \cdot 6 \cdot 8}\,x^4 + \dots,$$

which is the special case for the exponent $1/2$ of Newton's later Binomial Theorem.

The importance of such infinite series representations in the construction of tables is nicely expressed by George Biddell Airy (1801 - 1892), England's 7th Astronomer Royal from 1835 to 1881, in his *A Treatise on Trigonometry*[38], wherein, in the last chapter titled "On the construction of trigonometrical tables", we read:

(192.) But the natural sines for these arcs, at least for $10°, 20°$, &c., or more conveniently for $9°, 18°$, &c., may be calculated independently thus. We found for $\sin x$ the series

$$x - \frac{x^3}{1.2.3} + \frac{x^5}{1.2.3.4.5} - \&c.,$$

let $x = \frac{m}{n} \cdot \frac{\pi}{2}$; then $\frac{\pi}{2}$ being found by the differential calculus to $= 1,570796326794897$, we have $\sin \frac{m\pi}{2n} =$

$$\frac{m}{n} \times 1,570796326794897 - \frac{m^3}{n^3} \times 0,645964097506246$$
$$+ \frac{m^5}{n^5} \times 0,079692626246167 - \frac{m^7}{n^7} \times 0,004681754135319$$
$$+ \frac{m^9}{n^9} \times 0,000160441184787 - \frac{m^{11}}{n^{11}} \times 0,000003598843235$$
$$+ \frac{m^{13}}{n^{13}} \times 0,000000056921729 - \frac{m^{15}}{n^{15}} \times 0,000000000668804$$
$$+ \frac{m^{17}}{n^{17}} \times 0,000000000006067 - \frac{m^{19}}{n^{19}} \times 0,000000000000044$$

Similarly, as

$$\cos x = 1 - \frac{x^2}{1.2} + \frac{x^4}{1.2.3.4} - \&c. \text{ we have } \cos.\frac{m}{n}.\frac{\pi}{2} =$$

$$1,000000000000000 - \frac{m^2}{n^2} \times 1,233700550136170$$
$$+ \frac{m^4}{n^4} \times 0,253669507901048 - \frac{m^6}{n^6} \times 0,020863480763353$$
$$+ \frac{m^8}{n^8} \times 0,000919260274839 - \frac{m^{10}}{n^{10}} \times 0,000025202042373$$
$$+ \frac{m^{12}}{n^{12}} \times 0,000000471087478 - \frac{m^{14}}{n^{14}} \times 0,000000006386603$$
$$+ \frac{m^{16}}{n^{16}} \times 0,000000000065660 - \frac{m^{18}}{n^{18}} \times 0,000000000000529$$
$$+ \frac{m^{20}}{n^{20}} \times 0,000000000000003$$

The cosine of an arc being the sine of its complement, $\frac{m}{n}$ will never exceed $\frac{1}{2}$, and a few terms of these series will give the natural sines with great ease to 15 decimals.[39]

---

[38] The revised edition, which I downloaded from the Internet, was published in 1855.

[39] Airy, *A Treatise on Trigonometry*, revised edition, 1855, pp. 80 - 81. Airy implicitly assumes $0 \le m \le n$ so that

Note that Briggs did not use a power series expansion to calculate logarithms. It would take a few decades for power series to make their presence felt, chiefly through the work of Isaac Newton, who in the mid-1660's made the European discovery of the expansions of $\sin x$ and $\cos x$ cited above by Airy, and who discovered the expansion[40],

$$\ln(1+x) = x - \frac{x^2}{2} + \frac{x^3}{3} - \frac{x^4}{4} + \dots, \quad \text{for } -1 < x < 1,$$

and even used the series to calculate a small table of logarithms. The series diverges for $|x| > 1$ and cannot be used directly to find the logarithms of numbers greater than or equal to 2. Any number $y \geq 1$, however, can be written in the form

$$y = \frac{1+x}{1-x}$$

for $0 \leq x < 1$ and one can calculate

$$\ln y = \ln \frac{1+x}{1-x} = \ln(1+x) - \ln(1-x)$$

$$= 2\left(x + \frac{x^3}{3} + \frac{x^5}{5} + \dots\right) \tag{14}$$

3 Exercise. A conceptually simpler approach when $y \geq 2$ is to let

$$1 + x = \frac{1}{y}, \quad -1 < x < 0$$

and calculate

$$\ln y = -\ln \frac{1}{y} = -\ln(1+x) = -x + \frac{x^2}{2} - \frac{x^3}{3} + \frac{x^4}{4} - \dots \tag{15}$$

Calculate $\ln 10$ to 5 significant figures using each of (14) and (15). Which series would be more convenient for making a table?

The Calculus of Finite Differences would become a most important tool in the construction and use of tables in centuries to come. In its simplest manifestation, one wishes to construct a table of values $f(0), f(1), f(2), \dots$ of some

$$\frac{m\pi}{2n} \leq \frac{\pi}{2}.$$

If $m/n > 1/2$, one finds

$$\sin\left(\frac{m\pi}{2n}\right) = \cos\left(\frac{\pi}{2} - \frac{m\pi}{2n}\right) = \cos\left(\frac{n-m}{n} \cdot \frac{\pi}{2}\right),$$

with $(m-n)/n < 1/2$. Thus one can always carry out the computation using a fraction $\leq 1/2$.

[40] Newton was never one to publish quickly and the series is called *Mercator's series* after Nicolaus Mercator (*c.* 1619 - 1687) who was the first to publish it, which he did in his book *Logarithmotechnica* (1668). *Cf.* C.H. Edwards, Jr., *The Historical Development of the Calculus*, Springer-Verlag, New York, 1979, pp. 154 - 164 for details.

function $f(x)$. The differences between successive arguments are a constant 1. In Tables 2 and 3, above, we considered the differences in the values,

$$\Delta f(x) = f(x+1) - f(x),$$

and the differences of these differences,

$$\Delta^2 f(x) = \Delta f(x+1) - \Delta f(x).$$

One can also consider the third differences,

$$\Delta^3 f(x) = \Delta^2 f(x+1) - \Delta^2 f(x),$$

and the fourth, fifth, etc. Consideration of first differences and their use in linear interpolation goes back a long way. Second differences, or second order differences as they are perhaps more properly called, did not make so early an appearance. We have already seen an instance of their use in 12th century India by Bhāskara, above. Second differences and the corresponding interpolation formula had, in fact, been used in 6th or 7th century China in the work of Liú Zhuó (544 - 610). Third order differences were used by Wáng Xún (1235 - 1281) and Guō Shǒujìng (1231 - 1316) in the *Shòu shí lì* [*Works and days calendar*] (1280); and Zhū Shìjié went up to fourth order differences in his *Sìyuán yùjiàn* [*Precious mirror of the four elements*] (1303).[41] None of this was known in Europe at the time Briggs was applying Finite Difference methods.

Finite differences make their presence felt in tables in two ways. First, as we saw with Bhāskara and TABLE 3, above, we can use them to fill in the values of a table. Consider TABLE 5, below, in which we have tabulated the values,

x	0	1	2	3	$\dots$
$f(x)$	$f(0)$	$f(1)$	$f(2)$	$f(3)$	$\dots$
$\Delta f(x)$	$\Delta f(0)$	$\Delta f(1)$	$\Delta f(2)$	$\Delta f(3)$	$\dots$
$\vdots$					
$\Delta^n f(x)$	$\Delta^n f(0)$	$\Delta^n f(1)$	$\Delta^n f(2)$	$\Delta^n f(3)$	$\dots$

TABLE 5. Table of Differences

differences, second differences, and so on down to the n-th differences. If we know the first column and the last row, we can fill in the rest of the table by means of the simple recurrence relation,

$$\Delta^i f(j+1) = \Delta^i f(j) + \Delta^{i+1} f(j), \tag{16}$$

where we assume $\Delta^0 f = f$. That is, given the values of the j-th column, we obtain those of the $(j+1)$-st column from the bottom up, starting with the

---

[41] I refer the reader to Lǐ Yǎn and Dù Shíràn, (John N. Crossley and Anthony W.-C. Lun, translators), *Chinese Mathematics; A Concise History*, Oxford University Press, Oxford, 1987, for information on Chinese work with Finite Differences.

known $\Delta^n f(j+1)$ and generating $\Delta^{n-1} f(j+1), \Delta^{n-2} f(j+1), \ldots$ in turn using (16). Bhāskara did this centuries ago, and, in the mid-19th century, Airy discussed the method in some generality in his *A Treatise on Trigonometry*[42]. For trigonometric tables, the n-th differences can be calculated using the Addition Formulæ for Sines and Cosines. For polynomials the situation is even simpler: if $P(X)$ is a polynomial of degree n, the n-th difference $\Delta^n P(x)$ is a constant function. For example, if we extend TABLE 1 to include the differences, we obtain TABLE 6, below.

x	0	1	2	3	4	5
$f(x)$	0	1	4	9	16	25
$\Delta f(x)$	1	3	5	7	9	11
$\Delta^2 f(x)$	2	2	2	2	2	2

TABLE 6.

Notice that, if $\Delta^n f$ is constant, and one has $f(0), \Delta f(0), \ldots, \Delta^n f(0)$, the table can be extended as far as one wishes USING ONLY ADDITIONS, namely those demanded by (16). The importance of this observation extends beyond the tabulation of polynomials. For a nice smooth function, it often happens that on some given interval the k-th differences are very nearly constant. When this happens, one can tabulate values of the function in the given interval assuming a constant value for $\Delta^k f(x)$ and thus fill one's table using only easy to perform additions. Moreover, any nice smooth function can be approximated as closely as one wants on a given closed interval by a polynomial, hence if the error is smaller than the precision one is seeking, one can substitute the polynomial for the function in calculating one's tables.

A second use of the successive differences is in the problem of interpolation. First differences, as mentioned earlier, were often tabulated with the function values. Ptolemy did this in his chord tables, and the Indians followed suit. The inclusion of such information, readily found by subtraction, was a convenience in linear interpolation—finding the "part proportionall" as expressed in the title of Napier's *Descriptio* and Samuel Wright's preface thereto. Suppose one's table gave values $f(a)$ and $f(a+1)$ and one wanted the value of f at some point $a+h$ with $a < a+h < a+1$. The assumption that $f(a+h)$ divides the interval $[f(a), f(a+1)]$ in the same proportion as $a+h$ divides $[a, a+1]$,

$$\frac{f(a+h) - f(a)}{f(a+1) - f(a)} = \frac{a+h-a}{a+1-a} = \frac{h}{1} = h,$$

leads to the approximate value

$$f(a+h) \approx f(a) + h\big(f(a+1) - f(a)\big) = f(a) + h\Delta f(a). \qquad (17)$$

If, thus, one's table lists both $f(a)$ and $\Delta f(a)$, one has but to insert these values and h into this formula to perform the interpolation.

---

[42] Airy, *op. cit.*, pp. 81 - 83.

Using successive differences allows one to obtain possibly better approxima-tions to $f(a+h)$ than (17), which is linear in h. If one knows $f(a), \Delta f(a), \ldots, \Delta^n f(a)$, one has

$$f(a+h) \approx f(a) + h\Delta f(a)$$
$$+ \frac{h(h-1)}{2 \cdot 1}\Delta^2 f(a) + \ldots + \frac{h(h-1)\cdots(h-n+1)}{n(n-1)\cdots 1}\Delta^n f(a), \quad (18)$$

a formula often called *Newton's Forward Difference Formula* although Briggs used it in constructing logarithms and Thomas Harriot (*c.* 1560 - 1621), from whom Briggs may have learned it, had used it even earlier but had left it unpublished. For $h = 1, 2, \ldots, n$, the expression on the right actually agrees with f. For other values of h, it is only approximate.

To give us some idea of the use of Newton's Forward Difference Formula, I have written a little program for the *TI-83 Plus*. It asks the user to enter a list of values $f(0), f(1), \ldots, f(n)$ of some function f, finds the coefficients of (18), viewed as an n-th degree polynomial in h, and then asks for a value of h (called X in the program) and then calculates a value for $f(h)$:

```
PROGRAM:INTRPLTR
:Disp "PLEASE ENTER"
:Disp "A LIST."
:Input "L= ", L₁
:dim(L₁)→N
:N→dim(LDELTA)
:L₁(1)→LDELTA(1)
:For(I,1,N−1)
:ΔList(L₁)→L₁
:L₁(1)→LDELTA(I+1)
:End
:seq(X!,X,0,N−1)→LFACTS
:LDELTA/LFACTS→LDELTA
:N→dim(LXTERM)
:Disp "PLEASE ENTER"
:Input "X= ", X
:1→LXTERM(1)
:For(I,1,N−1)
:(X−I+1)∗LXTERM(I)→LXTERM(I+1)
:End
:sum(LDELTA∗LXTERM)→Y
:ClrList L₁
:DelVar N
:DelVar LDELTA
:DelVar LFACTS
:DelVar LXTERM)
```

To see how the extra terms affect the estimated value of $f(h)$, I recalled my comment somewhat earlier on how poor a result linear interpolation gave for finding $\sqrt{2.5}$ from my table in Chapter 9 of square roots:

$$\sqrt{2.5} \approx \sqrt{2} + \frac{\sqrt{3}-\sqrt{2}}{2} \approx 1.573132185.$$

In place of my table, I let the calculator find (admittedly) less accurate values quickly by entering

   "seq($\sqrt{}$(X),X,2,A+2)"→L$_2$.

This attaches the formula within the quotes to the list L$_2$. Updating the value n to be stored in A [43] will automatically update L$_2$. Then I successively stored the values 5, 10, 15, 20, 25, 50 in the variable A and ran the program INTRPLTR. When prompted for a list, I entered L$_2$ and, when prompted for a value X, I entered $h = 2.5 - 2 = .5$.[44] The results are collected in TABLE 7, below. The correct value is approximately 1.58113883 and we see the approximation

degree	approximation to $\sqrt{2.5}$
5	1.580973612
10	1.581115126
15	1.581131387
20	1.581135568
25	1.581137110
50	1.547844890

TABLE 7. Interpolants for $\sqrt{2.5}$

improving for a while (I have italicised the correct digits in the second column), even in the step from using a polynomial of degree 20 to that using one of degree 25, but the use of 51 terms and a polynomial of degree 50 gives the worst value of them all—even worse than a simple linear interpolation. Each additional term beyond the 25th has a product

$$\frac{.5(-.5)(-1.5)\cdots(.5-25+1)k_1}{k_2 \cdot 25 \cdot 24 \cdots 1},$$

where the numbers k_1 and k_2 themselves successively include more multiplications and divisions. Thus, there are at least 25 multiplications and 25 divisions per term, each with its own attendant rounding error. It would seem we have run once more up against the limitations of the calculator. But is that really the problem? Perhaps the slower growth of the function for larger values of x

---

[43] N would be a better name for the variable, but I have used N in the program and, variables being global, out of force of habit I chose a new variable name here. For no better reason than that A was the first real variable not used in the program, I chose it. In the present case, N is assigned the same value as A and there would have been no conflict. But it is good to get into the habit of playing it safe.
[44] Notice I start the sequence L$_2$ at $a = 2$.

and the occasional negative differences (e.g., Δ^2 is negative) have brought the value down.

Rewriting (18) as

$$f(a+h) \approx f(a) + h\Big(\Delta f(a) + \frac{h-1}{2}\big(\Delta^2 f(a) +$$

$$\ldots + \frac{h-2}{3}\big(\Delta^3 f(a) + \ldots + \frac{h-n+1}{n}\big(\Delta^n f(a)\big)\ldots\big), \quad (19)$$

one can reduce the total number of multiplications and thus the total number of possible rounding errors. In writing a new interpolation program based on this I've decided to make other improvements as well. Instead of asking for a list, the program assumes some function stored in Y_1 and generates a list therefrom. (The function stored could well be just a list: $Y_1(X) = L_2(X)$. So this is a generalisation, not a switch.) It then asks for an initial value a to be stored in the variable A, an increment Δx which it stores in the variable H (not to be confused with the h of (19). (INTRPLTR always assumes $\Delta x = 1$.) Then it asks for the degree n of the interpolation polynomial (which must be less than the length of L_2 if one has chosen $Y_1(X) = L_2(X)$), stores it in N, and calculates the first n differences of Y_1 at a. When this is done, it asks, as before, for an argument x for which to find the interpolated value. However, where before we entered $x - a$ for X, we now simply enter x and the program calculates $(x - a)/h$ and stores it in the variable X. I have also added some ClrHome commands to remove the clutter from the display screen and make its requests for input more readable:

```
PROGRAM:INTERPS
:ClrHome
:Disp "PLEASE ENTER"
:Disp "START NUMBER"
:Input "A=",A
:ClrHome
:Disp "PLEASE ENTER"
:Disp "INCREMENT"
:Input "H=",H
:ClrHome
:Disp "PLEASE ENTER"
:Disp "LEVEL OF"
:Disp "INTERPOLATION:"
:Input "N=",N
:seq(Y₁(X),X,A,A+N∗H,H)→L₁
:N+1→dim(ʟDELTA)
:L₁(1)→ʟDELTA(1)
:For(I,1,N)
:ΔList(L₁)→L₁
:L₁(1)→ʟDELTA(I+1)
:End
:ClrHome
```

```
:Disp "PLEASE ENTER"
:Input "X=",X
:(X−A)/H→X
:If N=1
:Then
:∟DELTA(1)+X∗∟DELTA(2)→Y
:Goto 1
:End
:augment({1},seq(X−I+1,I,1,N))→∟TERMS
:∟DELTA(N+1)→Y
:For(I,N,1,−1)
:∟DELTA(I)+∟TERMS(I+1)/I∗Y→Y
:End
:Lbl 1
:Disp Y
:ClrList L₁
:DelVar A
:DelVar H
:DelVar I
:DelVar N
:DelVar X
:DelVar Y
:DelVar ∟DELTA
:DelVar ∟TERMS
```

Running the program for $Y_1(X) = \sqrt{(X)}$, A = 2, X = 2.5, H=1 and N = 5, 10, 15, 20, 25, 50, respectively, yields the same values as INTRPLTR does. This strongly suggests the bad result for N = 50 is not due to round off errors. The only way to be certain, of course, is to carry out the computation with greater precision.

4 Project. (For SCHEMErs.) In SCHEME modify square.root to calculate the square root of an integer to, say 10 decimal places and represent the output as a rational number:

$$\frac{\text{integer}}{10^{10}}.$$

Create a list roots of such roots for $n = 2, 3, \ldots, 52$. Write a SCHEME program interpolator to calculate the terms on the right side of (18) *exactly*. Then run the program on roots for $h = .5$. Perform the long division[45] to 9 decimal places and compare the result with $\sqrt{2.5}$. Which of the two proposed explanations for the disastrous value of the last entry in TABLE 7 do you now favour?

Our story shall now take a different turn. All the major mathematical elements of table making were now in place. Mathematically speaking, the rest of the story is one of refinement—more decimal point accuracy, more entries in the tables, more functions to be tabulated, and, for the use of tables, new

---

[45] You might want to write a long division program in SCHEME to do this for you.

methods of interpolation. A great many names involved in this enterprise could be cited. One in particular is that of Jurij Veha (1756 - 1802), better known under the Germanised form of his name Georg von Vega. Vega was a comparatively minor figure in mathematics and appears in neither the *Dictionary of Scientific Biography* nor the *World Who's Who in Science*[46], but although not celebrated much today, Vega was honoured in his own day by being made an hereditary baron in 1800. Two years later, his body was found in the Donau [Danube]; he had been murdered by a miller.

Vega published 4 volumes of lectures on mathematics in 1782, 1784, 1788, and 1800, respectively. From 1783 on he published tables of logarithms, most famously his *Logarithmisch-trigonometrisches Handbuch* of 1793 and the *Thesaurus logarithmorum completus* of 1794. The former went through many editions. The 40th through 46th editions (1856 - 1862) were edited by Carl Bremiker (1804 - 1877). An 81st edition came out in 1906, and a quick visit to Amazon.com reveals a reprinting as recent as 1981.

Less important for the traditional history of mathematics, but of genuine importance for the social history of mathematics was a project almost contemporary, but slightly later than Vega's logarithmic publications. This was Revolutionary France's grand scheme to produce the mother of all tables.[47] Gaspard Clair François Marie Riche de Prony (1755 - 1839) led the project, which was to make the most complete and accurate collection of logarithmic and trigonometric tables.

Up to this point, tables had been either the work of individuals or, as in the cases of Rheticus and Otho, an individual and a small staff of expert assistants. De Prony had read Adam Smith's *A Treatise on the Wealth of Nations* (1776),

---

[46] Allen G. Debus, ed., *World Who's Who in Science; From Antiquity to the Present*, Marquis Who's Who Inc., Chicago, 1968. This massive single volume listing contains mini-biographies of around 30000 scientists. The information was gathered by a large staff of scholars, many of whom were still students. The result is not as reliable as the *Dictionary of Scientific Biography*, but with 5 or 6 times as many entries it is still valuable to have on one's bookshelf. The first important work of this kind was J.C. Poggendorff, *Biographisch-literarisches Handwörterbuch der exacten Wissenschaften*, published in two volumes in 1863. The tradition was continued under various new author/editors under variants of the title beginning in 1898 with *J.C. Poggendorffs Biographisch-literarisches Handwörterbuch zur Geschichte der exacten Wissenschaften*. An indication of the importance of this work is given by the fact that, following the end of the Second World War, the US Attorney General re-assigned the copyright to an American publisher and a reprint was issued in 1945.

[47] Oddly enough, the article on de Prony in the *Dictionary of Scientific Biography* doesn't discuss this. There is information on it in the articles on Jean Baptiste Joseph Delambre (1749 - 1822) and Legendre, but the most detailed accessible accounts are by the historian Ivor Grattan-Guinness: "Work for the hairdressers: the production of de Prony's logarithms and trigonometric tables", *Annals of the History of Computing* 12 (1990), pp. 177 - 185, and, a nicely illustrated revision thereof, "The computation factory: de Prony's project for making tables in the 1790s", in: Martin Campbell-Kelly, *et. al., op. cit.*

and changed that, transforming the art of table making into an assembly line using the cheapest labour possible—hairdressers rendered jobless by the French Revolution, retrained to perform simple arithmetic operations. He had a top tier of mathematicians, including Legendre and Lazare Nicolas Marguérite Carnot (1753 - 1823), who chose the formulæ to be used in calculating and checking the values, a second tier of "calculators", among whom Antoine Parseval (1755 - 1836), who calculated initial values and differences and prepared a set of tables to be filled in by the lowest tier—the retrained hairdressers who only had to perform additions and subtractions. For quality control, two sets of equations were used by two groups and the results were compared.

Although completed in 1801, De Prony's tables were never published. It was simply too expensive. Only two copies were made, each running to 19 volumes, 18 of tables and one giving an account of the methods used. The tables of logarithms of numbers from 1 to 200000 calculated to 14 decimal places occupy the first 8 volumes. There followed the numerous volumes of trigonometric functions. Negotiations with the English to publish the edition jointly to defray publication costs fell through because of the British preference for a circle of 360 degrees, each degree of 60 minutes, and each minute of 60 seconds: the revolutionary French division of the circle into quadrants of 100 degrees, each of 100 minutes, each of 100 seconds, did not please them

De Prony's project may have collapsed under its own weight, but the methodology caught on. Indeed, the two world wars of the 20th century resulted in massive war-related computational projects, most famously the production of ballistic tables by "computers", as the people, mostly young women, who performed the actual computations were called. In America, Oswald Veblen (1880 - 1960) was involved in making such tables during the First World War, and the story of the making of artillery tables in the Second World War[48] can be found in any book on the history of computers.

A harbinger of the decline in the perceivable importance of tables came with Charles Babbage (1791 - 1871), who reasoned that since tables could be constructed using only additions and subtractions the entire process could be mechanised, and set out in the 1820's to carry out the mechanisation. For various reasons he never did complete a full scale *difference engine*, as he called his envisioned tabular machine, but others following his lead did so and in 1991 the London Science Museum celebrated his 200th birthday by constructing a working difference engine built according to Babbage's specifications laid out in the late 1840's using only materials available in the mid-1800's. No tables of great importance were calculated by any of the various difference engines, but the "proof of concept" was established.

The modern computer has several origins. The concept was clearly enunciated by Babbage in the 1830's, although, as with his difference engine, he never completed what he called his *analytical engine*. A century later the computer was undergoing multiple births, most central to our story beginning in

---

[48] The tables of the First World War were rendered obsolete by the manufacture of more powerful guns: air resistance is lower at the higher altitudes newly achieved by the projectiles, and shells were travelling greater distances than predicted.

the 1930's in Iowa where John Vincent Atanasoff (1903 - 1995), mindful of many hours spent solving systems of linear equations in connexion with his dissertation in physics, designed a computing device for just this purpose. He was visited in 1941 by John William Mauchly (1907 - 1980), another physicist keen on electronic computation. When the United States entered the Second World War, while other physicists were applying their expertise to developing radar and nuclear bombs, Mauchly proposed to help the war effort by building an electronic computer. There, stealing[49] ideas from Atanasoff and others, he eventually completed construction of a working general purpose digital computer which had been christened ENIAC for *E*lectronic *N*umerical *I*ntegrator *a*nd *C*omputer. Although it was developed jointly by the Ballistics Research Laboratory of the Aberdeen Proving Grounds in Maryland and the Moore School of Electrical Engineering at the University of Pennsylvania, it was completed too late to be of use in the ballistics calculations being carried on by the WACs in Aberdeen. Its first calculations were on some top secret problem connected with the atom bomb project in Los Alamos in the summer of 1945. The computer itself was finished at the end of 1945 after the war's completion.

> The Army agreed to allow university scientists to use the ENIAC free of charge, and a number of problems were run under this arrangement. Profs. Hans Rademacher and Harry Huskey did computations of tables of sines and cosines to study the way round-off errors develop in numerical calculations (15 - 18 April 1946).[50]

The original electronic computers were large, expensive, few and far between, and, although faster than human calculators, slow by today's standards. They did not replace tables for several reasons. In this day of portable laptop computers with batteries of usable life, it may be hard to imagine that tables, which we might regard today as bulky, were more portable, more affordable, and easier to use: computers had to be programmed in very low level languages to perform any tasks. In my collection I have a textbook on programming published in 1957[51]. It teaches programming in machine and assembly language. The preface gives an idea of how widespread computers were not:

> Those who have a computer at hand while they read the book will find several desirable features. Possibly the most important is that the chapters which do not apply to a particular situation can be omitted without loss of continuity: at least half of the chapters may be omitted or included at will. This group of readers will find little difficulty in applying the illustrations to their particular machine, partly because of

---

[49] To be accurate, the theft occurred after war's end when he patented many of the inventions of Atanasoff and of his co-workers.

[50] Herman H. Goldstine, *The Computer: From Pascal to von Neumann*, Princeton University Press, Princeton, 1972, p. 232. The passage continues with other applications, but I thought this one particularly relevant to our present discussion.

[51] D.D. McCracken, *Digital Computer Programming*, John Wiley & Sons, Inc., New York and Chapman & Hall, Ltd., London, 1957.

the format of the programs and partly because TYDAC is an uncompli-
cated machine. In a classroom situation, the instructor can fairly easily
rewrite the illustration. The many exercises are in no way dependent
on the features of TYDAC.
Both groups of readers will find that the text is self-contained. If necessary, it may be read without an instructor or reference material, either
to provide a general background knowledge of computer programming
or as a supplement to a manual.[52]

In the early years, each computer was unique, with its own architecture and its
own language, termed machine language. This meant that a general introductory textbook had to be written for a specific, yet not too specific, machine. The
author created his own imaginary machine he called TYDAC (*T*ypical *D*igital
*A*utomatic *C*omputer). Not all schools at which programming was taught had
computers. And those that had computers may have had computers with special features or the parts may have had different parameters.

The present trend is for the main memory to be built around magnetic
cores in large machines, and magnetic drums in the smaller. Auxiliary
memory is almost always magnetic tape, with magnetic drums also
being used in the large computers. Electrostatic and mercury-delay
storage are still employed in some machines, but are being superseded
with newer ones.[53]

The size, say, of the magnetic drum will affect how much memory it has, thus
the range of addresses that can be used in writing programs. The speed of
rotation, combined with the speed of the computer's CPU may dictate where,
on a specific machine, one wants to store the results of a computation: pick the
wrong address and you have to wait almost a complete rotation before storing
it.

Chapter 1, section 4 of the text includes the following definition:

CODING
This is the writing of detailed machine instructions which carry out
the arithmetic operations called for above, whether in actual machine
language or in some symbolic or abstract form...It is usually the first
subject a newcomer to computing is taught. For most problems it is not
the most demanding aspect of the work, but in others involving severe
space or time restrictions it is crucial. It is the part of the job which
really comes face to face with the details, not to say peculiarities, of
the machine being used. Coding must to a certain extent be relearned
in order to work on a new computer, although of course relearning will
be much shorter than the original learning. It involves a great deal of
detailed work and is the source of many errors.[54]

The chapter ends with the following bit of conventional wisdom:

---

[52] *Ibid.*, p. vi.
[53] *Ibid.*, p. 3.
[54] *Ibid.*, p. 11.

It is almost never practical to set up a calculation for computer solution if the problem is to be done only once or only a few times.[55]

The point is that programming was hard. It could well have been easier to solve the problem by hand than to take the time writing a program to have the machine do the work.[56]

It may seem like I am getting off-track, but I am not really doing so. Computers were not widely available, each had its own language, and, something I've not explained yet, this language was limited. One dealt with 0's and 1's. Each base 10 digit was a string of 0's and 1's stored in a register at some address. The book even offers an entire chapter on "Decimal Point Location Methods". It would thus make sense to write a program to generate a table of, say, sines, but probably not to do so merely to calculate a single value of the sine function.

Speaking of tables, in Chapter 17, section 4, we find

TABLE LOOK-UP
In some applications and using certain equipment, the best way may be to store a table of values of the function at certain values of the argument—which is also in memory. Usually interpolation is used, either linear or a higher order formula. It may be feasible to store entries at closer spaced values of the argument where the function is changing rapidly. The locating of the values in the table which surround the argument of current interest may be done by any of the methods of section 17.1—which now appears in a more general light.[57]

I recall around 1980 visiting a major physics lab and being told my one of the researchers that they had performed many computations in advance and stored the results in a table on the computer. But computers are a lot faster these days and most functions can be calculated so quickly by easy-to-write programs in higher order languages that there is probably little need for such a practice on the computer. Tables, such as those we gave of the Fibonacci numbers or of the solutions to the Pell equation, still have their uses, e.g., in surveying the results to spot patterns or the lack thereof; but as tools for computation they have been rendered largely obsolete by the ubiquity of affordable, high-speed computers and high order computer languages.

Back in 1957, the situation was different and our cherished textbook concludes with 11 pages of tables converting numbers in octal (base 8) to decimal, a task easily performed on most pocket calculators.

---

[55] *Ibid.*, p. 12.

[56] It would probably have taken less time for me to have found $\sqrt{2}$ to 50 decimals by hand than it took me to write **SQROOT** and remove all the bugs. On the other hand, I would not have been certain of the result without having programmed the calculator to then multiply my solution by itself, and it would have been a lot less fun. Also, we would not have had the demonstration of the capabilities of the calculator.

[57] *Ibid.*, p. 206.

I got my own introduction to computers about a decade later with FOR-
TRAN II-D. Computers were no longer rare, but they were mainframes, and
were expensive and had to be stored in temperature-controlled rooms. The programming student at the university went to a room with a punch-card machine
and a card printer. He punched his cards, printed them out and debugged his
program. When he was satisfied everything was in order, he would go to a special window and turn his cards in and be told when to come back for the results.
When he came back he would be handed a stack of cards which contained the
program and its output. He would then run it through the printer and read
his results. While I imagine access was more direct in industry, it is hard to
imagine the installation of such machines in some computationally intensive
areas. There is no way an æroplane could accommodate such a computer for
use by the navigator, and I don't imagine too many ships had such computing
facilities for their navigators either.

Tables, not computers, remained the computational aid in everyday matters.
Computers could be used to generate tables—even, as I did in Chapter 9, and,
as originally envisioned by Babbage, in formatting and printing tables:

COMPUTER PROGRAM
The tables were generated on a Burroughs 220 computer at Smith,
Kline & French Laboratories by an iterative technique described below.
The computations were carried to 18 decimal places by an interpreta-
tive double-precision floating decimal routine. The computer program
edited the results, rounding to ten places, and a special plugboard was
wired to be used by the high-speed printer. For maximum clarity the
printer was kept at its minimum speed of 625 lines per minute, although
it is capable of handling print at the rate of 1500 lpm.[58]

It is perhaps worth adding that the computer allowed a reversal of practice.
Since Napier, one had tried to avoid multiplication, the most recent practice
in pre-computer days being to reduce the operations involved in making a
table to additions and subtractions wherever possible and to assign the actual
arithmetic computations to unskilled labour. Now, suddenly multiplication was
no problem. The program in question computed a function $e(n, r, p)$ using the
recursion

$$e(n, r + 1, p) = e(n, r, p) \cdot \frac{n - r}{r + 1} \cdot \frac{p}{q}, \tag{20}$$

where $q = 1 - p$. For given n and p, the program computed $e(n, r, p)$ successively
for $r = 1, 2, \ldots, n$. The first step was to calculate p/q:

Hence, each new value of r could be found with only 2 multiplications
and 1 division.[59]

The mid-1960's did see an interesting change. Engineering students at colleges, long recognisable by the slide rules dangling from their belts, began to

---

[58] Sol Weintraub, TABLES OF THE CUMULATIVE BINOMIAL PROBABILITY DISTRIBU-
TION FOR SMALL VALUES OF p, Macmillan Company, New York, 1963, p. xv.
[59] Ibid., p. xvi.

sport smaller attachments as they migrated to pocket calculators. The earliest calculators were *four function calculators*, so-called because they only calculate the four basic functions of addition, subtraction, multiplication, and division. But for their not having a print out, these were superior to the old business adding machines. And, although one armed only with a four-function calculator might look longingly and enviously at the newer models that were coming out, one had a powerful tool in one's hands. In the preface to an up-to-date book on the use of calculators in science and engineering that was published in 1975, one reads

> When the right numerical methods are used, the electronic pocket calculator becomes a very powerful computing instrument. "Micronumerical methods" that will help the reader to derive the most computing capability for every dollar he has spent on his pocket calculator are discussed here.
> Most of the methods work on *any* calculator.[60]

And later,

> *Part II* presents numerical methods and formulas for numerically evaluating advanced mathematical functions. It also deals with the nested parenthetical form of the most frequently used functions in advanced engineering mathematics. It is the nesting of a sequence of arithmetic operations in parenthetical form that is the basis for performing advanced analysis on the pocket calculator. For example, 14 multiplies, 2 divides, 2 sums, and 108 data entries, totaling 126 keystrokes and 5 data storage records, are needed for a three-digit floating-point evaluation of $\sin(x) \approx x - x^3/3! + x^5/5!$. But only 54 key strokes and *no data storage records* (on a scratch pad) are needed to evaluate $\sin(x) \approx x(1 - (x^2/6)(1 - x^2/20))$ to the same accuracy.[61] Though we would evaluate $\sin(x)$ in this manner only on a four-function calculator, this example does illustrate the point that many complex formulas usually requiring calculator memory to be numerically evaluted can be written in a "nested" form not requiring calculator memory and thus can be evaluated conveniently on even the simplest four-function pocket calculator.[62]

And a little farther down the page:

> Here, again, the emphasis is on making even the simplest four-function calculator capable of doing sophisticated analysis without memory. Such concepts as nested parenthetical forms and recursion formulas, when combined with those of Chebyshev economization and rational polynomial approximation, provide tremendous flexibility and accuracy in the numerical evaluation of even the most complex functions on the

---

[60] Jon M. Smith, *Scientific Analysis on the Pocket Calculator*, John Wiley & Sons, New York, 1975, p. vii.

[61] This is the main difference between my programs INTRPLTR and INTERPS.

[62] Smith, *op. cit.*, pp. viii - ix.

simplest four-function pocket calculator. In fact, the serious analyst can perform precision calculations unheard of until a few years ago—in the comfort and convenience of his home or while traveling on the job.[63]

Does this spell the beginning of the end for mathematical tables? Our æroplane navigator who couldn't bring a mainframe on board and scarcely had room for de Prony's 19 volumes of logarithmic-trigonometric tables could easily fit a calculator in his pocket. . .

Logarithmic tables survived at least until the end of the twentieth century.[64] A serious search might reveal later publications, but tables of logarithms and trigonometric functions are no longer necessary for the end user—at least not directly: Vast amounts of memory can be fit into small devices and great speed is now attainable, so most functions can be calculated from scratch on a computer and probably also on calculators, if not now, in the foreseeable future. Perhaps it reflects more on my lack of online research skills, but I found no definite statement on whether or not specific calculators used look-up tables in calculating specific functions[65], but there was an awful lot of information available on implementing look-up tables, so if seems tables are still around, but just less visible—unless, of course, you recognise modern spreadsheets as their evolutionary replacements.

It is some time since we actually did some mathematics, so I feel I should finish off the chapter with an impressive exercise or project. I am afraid though that I am rather lacking in imagination as all I can think of is to ask the reader to create a table. He might try following Smith in using a nested version of the three term polynomial approximation to the sine function to generate a small table of sines and then add more entries using INTERPS. Or, better yet, the reader might want to move the program over to the computer and automate the entire procedure—generate an initial table, use interpolation to fill in the gaps, and then print out a table, perhaps a simple comma-delimited file that can be imported into a spreadsheet. This, of course, goes way beyond what is appropriate for middle school, the initial impulse behind this book, but it does touch on nearly everything discussed in this chapter.

---

[63] *Ibid.*, p. ix.

[64] Amazon.com and Amazon.de list several as late as the 1980's and I find *Shaum's Mathematical Handbook of Formulas and Tables*, published in 1998, at Amazon.com, and *Schülkes Tafeln*, published in 2000, at Amazon.de. Several historical reprints after 2000 are also listed, but these were probably published for their historical interest rather than their computational use.

[65] Tables, of course, save time by taking up a lot of space. In 1959, Jack E. Volder published ("The CORDIC trigonometric computing technique", *IRE Transactions on Electronic Computers*, v. EC-8 (1959), pp. 330 - 334) an efficient new computational scheme that involves only the storage of relatively small tables, additions, subtractions, decimal shifts, and branching commands. Called the CORDIC scheme (for *Co*ordinate *R*otation *Di*gital *C*omputer), within a decade or two it became the basis for function evaluation on pocket calculators.

Index

www.ingramcontent.com/pod-product-compliance
Lightning Source LLC
Chambersburg PA
CBHW071651200326
41519CB00012BA/2475